Wild
ORCHIDS
of Scotland

ROYAL BOTANIC GARDEN EDINBURGH

Wild ORCHIDS of Scotland

BRIAN ALLAN AND PATRICK WOODS
WITH PHOTOGRAPHS BY SIDNEY CLARKE

EDITED BY NORMA M GREGORY
AND MARY BATES

EDINBURGH : HMSO

© Crown Copyright 1993
First Published 1993
Reprinted 1993

ISBN 0 11 494246 3

Designed by George Bowie
HMSO Graphic Design, Edinburgh

Typesetting by Davidson Van Breugel, Glasgow

Origination by Centre Graphics Limited, Livingston

Printed in Scotland for HMSO by J Thomson Colour Printers Ltd., Glasgow

Dd 287831 C20 12/93

HMSO publications are available from:

HMSO Bookshops
71 Lothian Road, Edinburgh, EH3 9AZ
031-228 4181 Fax 031-229 2734
49 High Holborn, London, WC1V 6HB
071-873 0011 Fax 071-873 8200 (counter service only)
258 Broad Street, Birmingham, B1 2HE
021-643 3740 Fax 021-643 6510
33 Wine Street, Bristol, BS1 2BQ
0272 264306 Fax 0272 294515
9-21 Princess Street, Manchester, M60 8AS
061-834 7201 Fax 061-833 0634
16 Arthur Street, Belfast, BT1 4GD
0232 238451 Fax 0232 235401

HMSO Publications Centre
(Mail, fax and telephone orders only)
PO Box 276, London, SW8 5DT
Telephone orders 071-873 9090
General enquiries 071-873 0011
(queuing system in operation for both numbers)
Fax orders 071-873 8200

HMSO's Accredited Agents
(see Yellow Pages)
and through good booksellers

Foreword

'Orchideae are remarkable for the bizarre figure of their multiform flower which sometimes represents an insect, sometimes a helmet with the visor up, and sometimes a grinning monkey: so various are these forms, so numerous their colours, and so complicated their combinations, that there is scarcely a common reptile or insect to which some of them have not been likened'.

John Lindley's words from An Introduction to the Natural System of Botany of 1830 graphically describe the range of variation in this intriguing group of plants. The interpretation of the complicated floral structure of orchids has occupied botanists for generations and the study of variation between and within species continues to challenge the taxonomist. The physiology and ecology of orchids are equally demanding of study: the former, most notably, for the little-understood dependence of orchids on mycorrhiza; the latter for the restriction of many species to very specialised habitats.

In addition to their fascination for the scientist, orchids have a universal aesthetic appeal – indeed, many would consider them among our most attractive wild plants. To accommodate both viewpoints this book deliberately unites in its authorship and content the scientific expertise of Patrick Woods, Brian Allan, Phil Lusby and Roy Watling with the photographic and artistic skills of Sidney Clarke and Mary Bates.

The combination of beauty, interest and rarity makes orchids a particularly precious and vulnerable element of the Scottish flora. While this book was in its final stages of preparation, the wreck of the oil tanker Braer off Shetland demonstrated with horrifying immediacy just how threatened, for example, our coastal wildlife is – and several of our rare orchids are confined to littoral/coastal habitats. Less dramatic but no less damaging in the long term are the pressures of farming, forestry and recreation on other orchid refuges, such as those in the uplands. Orchids have a long history in folklore and folk medicine – we must take great care lest our generation oversees the extinction of all but the commonest species.

The authors and I intend this book to contribute to the conservation of our Scottish orchids by making their identification easier, their biology and ecology more widely understood and the intricate beauty of their flowers more readily appreciated. We hope that in so doing we can add to the enjoyment, by all who love plants, of this rare and attractive family.

Dactylorhiza majalis subsp. *scotica* by Mary Bates. Reproduced by courtesy of Royal Mail.

DAVID S. INGRAM
Regius Keeper, Royal Botanic Garden Edinburgh

ACKNOWLEDGEMENTS

Some four years have elapsed between this book's conception and its appearance. During that time many people have contributed in one way or another to our knowledge and understanding of Scottish orchids. Our special thanks are due to Drs J. Dickson and A. J. Richards for their assistance with *Epipactis*, to Dr F. Rose with *Gymnadenia*, to D. J. Tennant and Dr M. R. Lowe for valuable advice on the difficult genus *Dactylorhiza* and to F. Horsman for help with *Spiranthes*. We are grateful to F. Horsman, D. S. Lang, R. Piper, Dr A. J. Richards, Dr F. Rose, D. J. Tennant and D. M. Turner Ettlinger for their constructive comments on the key. Invaluable information was provided by the Botanical Society of the British Isles vice-county recorders. We thank the following who have assisted in one way or another: Mrs J. H. Allan, C. Badenoch, M. E. Braithwaite, A. Buckam, Dr J. K. Butler, Lt. Com. and Mrs H. D. Campbell-Gibson, A. Currie, Dr R. E. C. Ferreira, D. Grant, Professor D. M. Henderson, Dr M. Hughes, D. Hunt, Dr C. Jeffries, Mrs M. R. Maan, Mrs C. W. Murray, Miss R. Neiland, A. Panter, R. Payne, Dr C. D. Preston, Dr J. Roberts, F. Robertson, J. Grant Roger, M. N. Russell, A. J. Smith, K. Stevenson, T. Stevenson, K. Watson, Miss L. M. Watson, C. Wilcock, Mr & Mrs G. Wiltshire, Miss J. Wright and the many other friends and acquaintances who helped us. We also thank especially our colleagues here at the Royal Botanic Garden Edinburgh for their interest and enthusiasm.

Frontispiece A clump of the almost white-flowered form of *Dactylorhiza fuchsii* subsp. *fuchsii* with unspotted leaves growing amongst rough grass is a spectacular sight. Argyll, 28 vi 1990.

Contents

1. ORCHID
Biology

Orchid Biology

Patrick Woods and Roy Watling

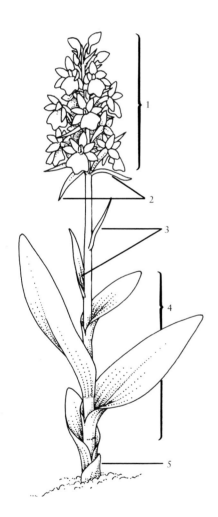

Parts of an orchid plant
1 inflorescence
2 bracts
3 non-sheathing leaves
4 sheathing leaves
5 sheath

Worldwide there are probably between 20,000 and 25,000 species of orchid. They comprise one of the largest and most diverse families in the plant kingdom with a huge range of forms occurring throughout all but the coldest and driest regions of the world. Only a few species are found within the Arctic circle and none in Antarctica; in deserts, they are confined to a few species in areas of higher humidity. Orchids reach their greatest number and diversity in the humid tropics where most species occur as epiphytes on the branches and trunks of trees.

Orchids fall into the class of flowering plants known as the monocotyledons which are characterised by the presence of one cotyledon (seedling leaf) at germination, leaves with parallel venation, and flower parts in threes or multiples of three. However, not all monocotyledons show all of these features: the majority of orchids for example have minute dust-like seeds which produce no cotyledon at all.

The absence of a seedling leaf and the lack of food reserves in the tiny seeds means that orchid seedlings are dependent on an outside food source. This is derived from an intimate association which develops between the young orchid's roots and a fungus, resulting in a dual organism known as a mycorrhiza. The majority of flowering plants are mycorrhizal to some degree. For orchids, the relationship is particularly important until green leaves are formed, but some species are dependent on mycorrhizas throughout their life. Much important research on orchid mycorrhizas has taken place in Scotland and so it is appropriate in a book on Scottish orchids to consider the relationship in some detail.

Orchid mycorrhizas are a particular kind known as endomycorrhiza, in which the fungal partner lives inside the root cells of the host. Orchid seed may start to germinate without the fungus but the process is completed only when the appropriate fungus has infected the embryo. It then swells, splits the seed coat, and forms a colourless protocorm which eventually develops chloroplasts and then shoot initials and roots. Although the fungus inhabits the root cells, parts of it remain outside the plant and are capable of breaking down cellulose and other resistant carbon compounds into soluble products which can be translocated into the growing orchid. Most orchids develop perennating storage organs, rhizomes or tuberoids, and the roots die off at the end of the season. The storage organ does not become infected by the fungus but, as new roots regrow the following season, they become colonised from the soil. Several years of build-up may be required before a flowering shoot is formed. Some *Dactylorhiza* and *Orchis* species may flower three or four years from germination but others take much longer. The common twayblade, *Listera ovata*, does not produce a leaf until the fourth year, and is said to take 13 to 15 years before flowering. It is not clear whether orchids continue to draw food from mycorrhiza once they become fully adult green plants. However several orchids lack chlorophyll. *Neottia nidus-avis* and *Corallorhiza trifida* are the only Scottish examples that cannot manufacture food for themselves and these remain totally dependent on mycorrhiza throughout their life.

The tuberoids of genera such as *Dactylorhiza* and *Orchis* die away when exhausted and a new one is formed each year. Food-storing rhizomes and roots eventually die back as new ones develop; orchids with this habit, such as *Epipactis*, *Listera* and *Neottia*, may produce several flowering stems growing close together. Uniquely among Scottish orchids, *Hammarbya paludosa*, the bog orchid, stores food in a pseudobulb, a swollen stem base hidden in the moss in which the plant grows. A pseudobulb is a type of storage organ commonly found in tropical orchids which stores water rather than food. The bog orchid produces a new pseudobulb every year, the old one gradually dying away.

Some orchids may flower for many years, whereas others, such as the fragrant orchid, *Gymnadenia conopsea*, are thought to flower only once or twice before dying. Apart from the creeping ladies' tresses, *Goodyera repens*, which is evergreen, the aerial parts of all Scottish orchids die back each year, although old dry fruiting stems occasionally persist.

Many orchids reproduce vegetatively, and the resulting plants develop much more rapidly than those from seed. In some species buds from roots or rhizomes form new plants when the connecting part dies away and, in those species that produce tuberoids, more than one bud may develop during a season, forming separate plants when the parent dies. The bog orchid is alone among Scottish species in that it produces tiny bulbils on the edges of the leaves; these detach and float away, quickly become infected by mycorrhiza, and form new plants.

Although orchids may vary tremendously in appearance from one genus to another, the flowers have certain structural features unique to the family which make it relatively easy, even for the layman, to recognise them as belonging to an orchid.

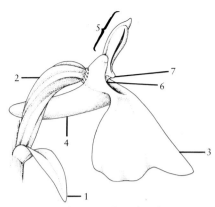

Dactylorhiza flower with sepals and two petals removed
1 bract
2 ovary
3 lip
4 spur
5 anther
6 stigma
7 viscidium

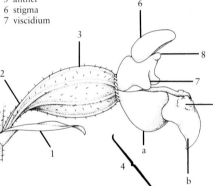

Epipactis flower with sepals and two petals removed
1 bract
2 pedicel
3 ovary
4 lip – a, hypochile, b, epichile
5 boss
6 anther cap
7 stigma
8 rostellum.

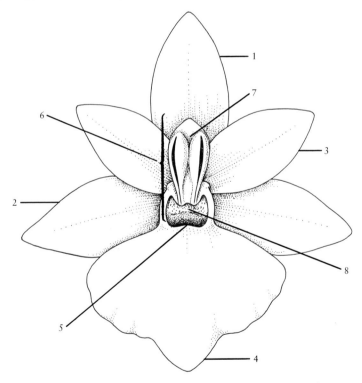

Left. Parts of a dactylorchid flower
1 dorsal sepal
2 lateral sepal
3 petal
4 lip
5 entrance to spur
6 column
7 anther
8 stigma

An orchid flower consists of a perianth of six segments arranged in an outer and an inner whorl at the top of the flower stalk (pedicel), which includes the ovary. The outer whorl comprises three sepals, which are more or less the same size, shape and colour. The uppermost sepal is referred to as the dorsal sepal, the other two are the lateral sepals. The inner whorl is made up of three petals of which two are alike and may be similar to the sepals. The third petal is almost always different from the other two in shape, size and colour being usually larger, sometimes lobed, often spurred and distinctly coloured or marked. It is known as the labellum or lip, and in *Cephalanthera* and *Epipactis* is divided into a hypochile and epichile. In most orchids, the lip is held in the lowermost position and points downwards. Such flowers are known as resupinate since the pedicel twists through 180° during development. Where the lip is uppermost, the flower may be non-resupinate (i.e. not twisted at all) or hyper-resupinate (twisted through 360°). All Scottish species have resupinate flowers with the exception of the bog orchid, which is hyper-resupinate. The simple explanation for the difference in shape, size and colour of the lip is that it is adapted to attract pollinators, often by presenting an inviting landing platform.

 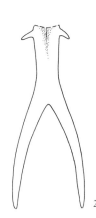 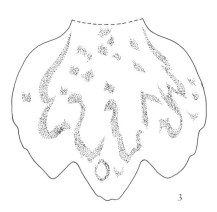 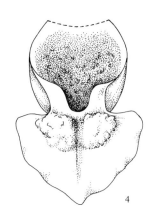

Lip types in Scottish orchids
1 strap-shaped
2 divided
3 3-lobed with honey guides
4 with hypochile and epichile

Other significant features of the flower are the highly modified reproductive organs. Instead of being separate, as in most other plant families, the male and female parts of an orchid flower are fused into a single structure known as the column which projects at an angle from the centre of the flower, to lie opposite the lip. In all Scottish genera the column consists of one fertile stamen with an anther at the top in which the pollen grains are aggregated into two or four often club-like masses called pollinia. The stigma is situated below the anther and is usually differentiated into a rostellum and a fertile stigmatic area. In some species the rostellum has a viscidium to which the pollinia are attached by stalks or caudicles. After pollination and fertilisation, the ovary develops into a capsule containing many thousands of seeds.

Cross-pollination between individuals of a species has the advantage over self-pollination of perpetuating genetic mix. In consequence, most species have evolved a variety of methods to ensure that pollen from the flowers of one plant is transferred to the stigmas of a different plant. The strategies by which cross-pollination is achieved make one of the most fascinating features of the orchid family, which has a whole range of ingenious modifications designed to attract the pollinator and ensure the successful transfer of pollen from one flower to another. Orchids have evolved a method of glueing their pollinia on to a visiting insect, often as it leaves the flower, by means of sticky exudates. In some genera, such as *Cephalanthera*, this is produced by the stigma but in most cases the rostellum is involved. In *Listera* the glue is under pressure and squirts out when the rostellum is touched. More commonly, there is a modified part of the rostellum, the viscidium, which detaches as a unit with the pollinia and sticks to the insect. Some genera, such as *Orchis* and *Dactylorhiza*, have a further modification in that the viscidium is covered by a sheath, the bursicle, which easily breaks when touched by an insect. The success of cross-pollination may be

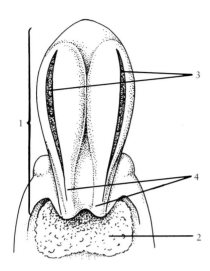

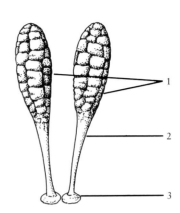

enhanced by changes that take place in the pollinia after they are detached. Thus, in *Anacamptis pyramidalis*, once the pollinia are stuck to the pollinating insect the angle at which they are held changes as the caudicles dry out to bring them more readily into contact with the stigma of the next flower visited. The same process makes the caudicles brittle so they are more easily transferred from the pollinator to the sticky stigma. There are of course many other variations and refinements to the process of pollen transfer some of which, particularly in Mediterranean and tropical species, may approach the bizarre.

In Scotland, pollinators of orchids include bees, wasps, hoverflies, flies, butterflies and moths, and to a lesser extent small beetles. In common with most insect-pollinated plants, orchids have a variety of mechanisms to attract their pollinators, although of course the insect visits the flowers entirely for its own ends. Distinctive colours, shapes and markings on the petals, especially the lip, serve as honey-guides that provide signals which the insect associates with the presence of nectar. Scent, often in combination with white flowers, is an effective signal to night-flying moths. The food-seeking insect is deceived by some orchids such as *Orchis* and *Dactylorhiza* which, despite providing the right signals, have no nectar. An insect will nevertheless visit and pollinate many flowers before learning better. Orchid pollen itself cannot be collected as food because it is aggregated into pollinia. A food reward is not the only attractant to pollinators offered by orchids. *Ophrys apifera*, recorded for southern Scotland in the *Flora of the British Isles* in 1952 but now thought to be extinct there, has flowers bearing an extraordinary resemblance to bees. The attractant is sex, and in attempting to mate with the flower, the insect picks up the sticky pollinia. Although there are many fascinating adaptations to cross-pollination in the Orchidaceae, self-pollination is normal in a few genera, *Epipactis* for example, and often occurs in others when cross-pollination fails.

Column of a dactylorchid
Left
1 anther
2 stigma
3 pollinia
4 viscidia covered by bursicle

Right
1 pollinia
2 caudicle
3 viscidium

Removal of pollinia from *Listera ovata* (note nectar on lip). Based on photographs from Baumann & Baumann, *Das Geheimnis der Orchideen* © Hoffmann und Campe Verlag, Hamburg 1988.

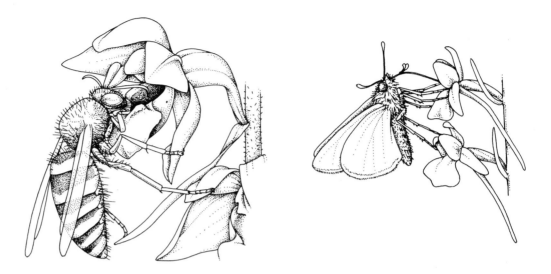

Left. Wasp on *Epipactis helleborine* (note pollinia attached to head). Based on photograph from Baumann & Baumann, *Das Geheimnis der Orchideen* © Hoffmann und Campe Verlag, Hamburg 1988.

Right. Skipper butterfly feeding on *Gymnadenia conopsea* (note pollina attached to proboscis). Based on photograph from Baumann & Baumann, *Das Geheimnis der Orchideen* © Hoffmann und Campe Verlag, Hamburg 1988.

The structure of an orchid flower is often related to the type of insect that effects pollination. The more specialised the flower structure and hence the more specific the pollinator, the less likely it is that hybridisation will occur, or that other insects can steal the nectar. Some part of the insect must come into close contact with the pollinia and so it must be lured to push into the flower to reach the nectar (or promise of nectar). For example, *Platanthera* flowers are white and night-scented with a downturned lip of little use as a landing platform, and a long nectar-containing spur. Accordingly, they are adapted to pollination by night-flying moths which hover to feed and inadvertently pick up the pollinia on their long proboses. Short-tongued insects cannot reach the nectar. *Anacamptis* and *Gymnadenia* flowers are brightly-coloured, scented, and have a broad lip as a landing platform and a long spur; they are pollinated by butterflies which are day-flying, must land to feed, and have a long proboscis, and sometimes by moths. Brightly coloured flowers with a prominent landing platform marked with lines and patterns have evolved to attract bees, although other insects also visit them. Flowers with shallower nectaries and spreading, rather dull flowers are pollinated by insects that have biting or short sucking mouthparts, such as wasps and flies. Wasps are the main pollinators of most *Epipactis* species, the nectar being secreted by the shallow hypochile of the lip. *Listera ovata* secretes nectar from a groove on the labellum and is pollinated by flies and ichneumon wasps. Beetles, spiders, ants and small flies all visit orchid flowers and occasionally effect pollination.

After fertilisation the fruit ripens and dries. The capsule splits longitudinally into six parts, three of which bear the numerous seeds. They are wind-dispersed and, being tiny, may be carried long distances. With germination and growth a long and hazardous process, it is necessary for many seeds to be produced for a few to reach maturity successfully.

Ripe seed capsule

Hybridisation among orchids is not unusual although some genera hybridise more readily than others; in Scotland the dactylorchids are prime examples. Both interspecific and intergeneric hybrids occur in the Orchidaceae though the former are more common. The normal convention for naming hybrids is whereby the parents are written in alphabetical order with a multiplication sign between. Where the hybrid has a specific or generic name of its own, the 'x' is placed before this as in *Dactylorhiza* x *venusta*, an interspecific hybrid, and x *Pseudadenia schweinfurthii*, an intergeneric hybrid.

Identification of the parents is often speculative in suspected hybrids. Difficulties arise when faced with hybrids which are themselves the product of repeated back-crossings between a fertile hybrid and one or other of its parents, or between the hybrids themselves. The range of plants thus produced is called a hybrid swarm, sharing a wide range of characteristics intermediate between both parents. If the parent plants are well adapted to their respective environments, then the hybrid plants may not thrive in either, and those most closely related to one or other of the parents will tend to survive. However, if there is an intermediate habitat, and these are often man-made, then more intermediate plants may survive in the hybrid swarm. Plants may show hybrid vigour, a phenomenon caused by the chance recombination of the most favourable genetic characters and resulting in offspring that are much taller or more robust than either parent. The reverse situation is equally likely but the resulting weaker and unfit plants die out rapidly. Distorted or abnormal plants of any species can occur and may be confused with hybrids.

Above left Dactylorhiza purpurella subsp. *purpurella* x *Dactylorhiza fuchsii* subsp. *fuchsii*. Roxburghshire, 2 vii 1989.

Above centre Epipactis youngiana x *Epipactis leptochila* subsp. *dunensis*. Lanarkshire, 3 viii 1990.

Above right Dactylorhiza fuchsii subsp. *fuchsii* x *Gymnadenia conopsea* subsp. *borealis*. Roxburghshire, 28 vi 1989.

There are several factors which govern the possibility of hybridisation. Plants must occur in the same area, their flowering times must overlap and both the parents must be visited by the same pollinators. Moreover, the floral anatomy of the parents must be sufficiently alike to enable the transfer of pollinia from one plant to the stigma of another. Even when all these conditions are fulfilled, pollen incompatibility may prevent fertilisation taking place. Interspecific hybrids are by far the most frequent, especially among the dactylorchids, as the floral anatomy is so similar and they have pollinators in common.

Although British orchids generally occupy rather specialised habitats, some can tolerate a variety of different conditions. Few species can be said to be common; many are rare and, as becomes apparent in the accounts that follow, some species fluctuate greatly in number. In trying to come to understand this fickleness in our orchid flora it is perhaps as well to recall that, in their early years at least, they are dual organisms. The ecology of the fungal partner also needs to be taken into account though unfortunately little is known about this. However, it is known that mycorrhizas are important in providing phosphates, the availability of which may be limiting in some habitats colonised by orchids. The fungi that can produce mycorrhizas are extremely varied but many of them are typical members of the microbial communities of rotting wood or dead herbaceous stems. Some may even be pathogens of other living plants. They are therefore active and important members of the ecosystem in their own right. We have yet to learn whether some of the reactions of orchid colonies to site changes may in fact reflect the behaviour of their fungal partners. Where known, the mycorrhizal associate of Scottish orchids is given in the species accounts.

At present much of our knowledge of mycorrhizas is based on laboratory study and extends only in a limited way to field conditions. Nevertheless we can at least use this to allow us to grow orchids under artificial conditions and so have a method by which rare species on the verge of extinction can be maintained.

Below *Dactylorhiza fuchsii* subsp. *hebridensis* (left) growing with *Coeloglossum viride*. Lewis, 12 vii 1991. The resulting hybrid is pictured opposite.

Opposite The intergeneric hybrid *Dactylorhiza fuchsii* subsp. *hebridensis* x *Coeloglossum viride*. Lewis, 12 vii 1991.

2. Orchid *Habitats and Conservation*

Orchid Habitats and Conservation

Phil Lusby

Orchids occur in a wide range of habitats in Scotland. However, there are no truly aquatic species and they are generally poorly represented at high altitudes although *Coeloglossum viride* can be found in wet mountain gullies. They are absent from salt-marshes but some species that occur near the coast on small islands, *Dactylorhiza fuchsii* subsp. *hebridensis* for example, must be tolerant of certain levels of salinity. Apart from these extreme conditions, orchids can be found in habitats that vary greatly in climate and soil conditions.

For most of the main habitats (grassland, heath, woodland and mire for example) there are recognisable differences between upland and lowland types as well as differences related to nutrient levels and pH, that is between basic, neutral and acid. However, there are no clear-cut boundaries. A useful and practical division between upland and lowland that works quite well in Scotland is the upper limit of enclosed land. Upland beyond the potential tree limit is termed montane and that below is sub-montane.

Boundaries between habitats are often ill-defined: orchids that are normally associated with grassland, for example, may occur in open areas within woodland. In the following short accounts of the main orchid habitats in Scotland the species most typical of each will be highlighted but this does not mean that other orchids may not occasionally occur, especially when transitions between habitats are diffuse. Orchids that are equally at home in more than one habitat will be mentioned in each appropriate section.

Natural basic grassland is a favourite habitat for *Orchis mascula*. Berwickshire, 13 v 1989.

Coniferous Woodland

Coniferous woodland in Scotland consists of native pinewood and forestry plantations. The distinction between the two is often clear but in old plantations of Scots pine the ground vegetation can sometimes be almost as rich as its native counterpart.

Native pinewood has become fragmented through the centuries into a number of scattered pockets but there are still quite extensive areas at Rothiemurchus and Abernethy on Speyside, the Mar estate on Deeside and around Beinn Eighe in Wester Ross.

The most characteristic orchid of the pinewoods is *Goodyera repens* which can occur in hundreds in some areas. The plant has also become well established in various plantations, and some of the older plantings in Culbin Forest, for example, carry very impressive colonies. *Corallorhiza trifida* and *Listera cordata* also frequently occur in pinewoods but are by no means confined to them. The canopy of native pinewoods does not necessarily consist solely of pine and may have areas where there is a mixture of pine and birch with rowan. *Listera cordata* can be equally frequent under birch and heather if the ground conditions are suitable. The very dense canopies of some of the non-native conifers used for forestry such as Sitka spruce and lodgepole pine are not very amenable to orchids or flowering plants generally, but *Goodyera repens* seems tolerant of very low light levels and has been found surviving in a non-flowering state in dense shade under lodgepole pine.

Coniferous woodland is the principal habitat for *Goodyera repens*. Roxburghshire, 22 vii 1989.

Deciduous Woodland

Through clearance and changes in land use, deciduous woodland has, like semi-natural coniferous woodland and other habitats, become fragmented. Large tracts of deciduous woodland are rare in Scotland but the dry birchwoods of central and eastern Scotland contrast with the damp oceanic woodlands of the Western Highlands and together constitute an internationally important habitat for flowering and non-flowering plants. Woodland of alder, willow and birch on very wet ground (carr woodland) is another distinct woodland type and occurs mainly in association with mires and along water courses. Natural beechwood, a rich orchid habitat in England, is absent from Scotland, although planted beechwoods occur widely.

Woods dominated by birch are not particularly rich in orchids but *Dactylorhiza maculata* subsp. *ericetorum* is common in less shaded areas. *Listera cordata* often occurs and where common imparts a northern 'flavour' to the woodland. On less acid soils in birch woods *Epipactis helleborine*, *Listera ovata* and *Dactylorhiza fuchsii* occur. In wetter birch woodlands *Corallorhiza trifida* may be found, but it is in the very wet birch-willow carrs that it appears in hundreds in some years.

On basic soils in oak-ash woodland (sometimes with planted beech) the attractive *Cephalanthera longifolia* may occur as a rare plant, mainly in the west. More widespread in this woodland type is *Epipactis helleborine*, *Orchis mascula*, *Platanthera bifolia*, *P. chlorantha*, *Dactylorhiza fuchsii*, *Listera ovata* and the saprophytic *Neottia nidus-avis*.

In river valleys on limestone in northern Scotland, where wych elm accompanies ash and oak, *Epipactis atrorubens* occasionally occurs although it is usually associated with more open limestone habitats such as rock crevices or grikes in limestone pavements.

Ancient, semi-natural woodland is more likely than other types to harbour rare plants but a notable exception is the mixed woodland that has colonised coal mine spoil-heaps (bings) in the industrial central belt of Scotland. Two very rare orchids have been discovered in this habitat, *Epipactis youngiana* and *Epipactis leptochila* var. *dunensis*. These are sometimes accompanied by the more common *Epipactis helleborine*.

Orchis mascula and *Primula vulgaris* thrive in open deciduous woodland. East Lothian, 20 v 1991.

Dactylorhiza fuchsii subsp. hebridensis can be found in vast numbers on the machair of the Western Isles. Harris, 17 vii 1991.

Grassland

Grassland is an extremely variable habitat. The floristic composition, and hence the appearance, varies widely between upland and lowland types and between those of different nutrient status.

Lowland grassland that has not been agriculturally improved is scarce in Scotland and most has undergone some degree of modification in recent years. Like most flowering herbs, orchids are outcompeted by grasses when grassland is fertilised and they do not tolerate herbicides. Obviously, it is in the least improved grasslands that most orchids survive. For this reason, upland grassland is generally richer in orchids than lowland – the lowland machair being an important exception – but few if any orchids are confined to the uplands. *Gymnadenia conopsea, Platanthera bifolia, P. chlorantha* and *Orchis mascula* all have a wide tolerance of soil conditions and occur in a range of grassland types. *Coeloglossum viride* has a similarly wide tolerance range and is one of the few orchids to extend into the montane zone. It does not survive in tall, rank grassland. The attractive *Pseudorchis albida* is more characteristic of upland grassland but, although it has been recorded at 600m in Argyll, is not generally a plant of high altitudes. *Dactylorhiza fuchsii* can be found in mildly acid grassland but is more at home in basic to neutral grassland. It is generally replaced by *Dactylorhiza maculata* subsp. *ericetorum* in acidic grassland but their tolerance ranges overlap. Hybrids have been recorded where the two occur together but are rarer than sometimes thought.

Machair is a distinctive grassland that occurs in oceanic areas and is generally associated with the Western Isles. Even though the greatest extent of this habitat occurs in the Outer Hebrides, the imprecision of the term means that machair is not unique to this region or even to Scotland. The criteria for a habitat to be described as machair are that it is an area of level or gently sloping, calcareous, shell-rich sand that supports short grassland without the long dune grasses such as marram (*Ammophila arenaria*). Machair has a history of grazing but more recently has been used for arable silage. The less agriculturally improved machairs are extremely colourful in the summer months with an abundance of flowering herbs. There are some orchids that are confined to this rich habitat, the best known being *Dactylorhiza fuchsii* subsp. *hebridensis* which can be seen in thousands in certain places in July. The recently recognised *D. majalis* subsp. *scotica* is currently known only from two sites on the machair of North Uist. Other species that occur more commonly include *Listera ovata, Dactylorhiza purpurella, D. incarnata* subsp. *coccinea, Coeloglossum viride, Gymnadenia conopsea* subsp. *borealis* and *Platanthera bifolia*. In 1989 and again in 1991 an unusual intergeneric hybrid between *Dactylorhiza fuchsii* subsp. *hebridensis* and *Coeloglossum viride* was found.

Unimproved neutral grassland supports orchids such as *Dactylorhiza fuchsii* subsp. *fuchsii* as well as an abundance of herbs such as *Lychnis flos-cuculi*. Midlothian, 6 vii 1990.

Dune slacks along the east coast are home to many orchid species. Fife, 8 vi 1989.

Margins of freshwater lochs on the west coast and Hebrides are one of the habitats of the rare *Spiranthes romanzoffiana*. Coll, 22 viii 1991.

Dune Slacks

Dune slacks are the low-lying hollows between stabilised dunes that can become flooded, especially in winter. They are a particular feature of the east coast of Scotland with good examples at Aberlady Bay south-east of Edinburgh, Tentsmuir in Fife, St Cyrus in Kincardine and the Sands of Forvie north of Aberdeen. Many of the orchids of the machair occur in dune slacks, including *Dactylorhiza incarnata* subsp. *coccinea*, *D. purpurella*, *Platanthera bifolia*, *Coeloglossum viride*, and a diminutive form of *Listera ovata*. *Corallorhiza trifida*, although not confined to dune slacks, is a characteristic species of the habitat and can occur in high numbers in some localities. *Epipactis palustris* has been recorded in the past from dune slacks at Aberlady Bay but does not seem to occur there now. The nearest locality to this former site is in slacks on Lindisfarne, Northumberland.

Marshes

Marshes are usually dominated by rushes and tall grasses. In habitat classifications marshes can be included with mires as non peat-forming types, or with grasslands as wet neutral types. Marshes in fields have suffered extensive drainage or have been, and continue to be, ploughed out. In less intensively farmed areas marshes often have a wide range of flowering herbs, and orchids such as *Dactylorhiza purpurella*, *D. incarnata* and *D. fuchsii* can be abundant. *Gymnadenia conopsea* subsp. *borealis* and both species of *Platanthera* occur, though less frequently. In a few western localities *Spiranthes romanzoffiana* occurs in marshes, especially where periodic flooding occurs, but it is also a plant of stony loch margins.

Calluna vulgaris and *Erica cinerea* are the dominant species of dry heathland. Coll, 24 viii 1991.

Mires, Fens and Heaths

Mires and heaths cover vast tracts of ground in upland Scotland. Mires or peatlands predominate in the north and west where the cool, wet weather provides conditions for peat formation on gently sloping ground, especially in hilly districts where rainfall is higher. In the drier east, peat formation is less pronounced. Mires on gentle slopes (blanket mire) can be dominated by a few plants including purple moor-grass (*Molinia caerulea*), heather (*Calluna vulgaris)*, deer-grass (*Trichophorum cespitosum*), cross-leaved heath (*Erica tetralix*) and the cotton grasses (*Eriophorum vaginatum* and *E. angustifolium*). Blanket mire is not a rich orchid habitat but *Dactylorhiza maculata* subsp. *ericetorum* and *Listera cordata* occur, and *Pseudorchis albida* and *Gymnadenia conopsea* subsp. *borealis* can sometimes be found growing in areas of purple moor-grass.

In wetter areas, such as in flushes where drainage water flows at or just beneath the surface, mosses and liverworts are often an important component of the vegetation. Here *Hammarbya paludosa*, usually only 2–10cm high, may be found but it is easily overlooked. When the ground water is calcareous, flushes are often richer and can include *Dactylorhiza incarnata*, *Platanthera bifolia* and *P. chlorantha*. Within the last ten years three marsh-orchids, all of which grow in basic or mildly basic flushes, have been added to the Scottish flora. These are *Dactylorhiza incarnata* subsp. *cruenta*, *D. lapponica* and *D. traunsteineri*. The first was previously known only from Western Ireland. *Dactylorhiza lapponica* is new to the British Isles (otherwise a Scandinavian plant) and *Dactylorhiza traunsteineri* was confirmed as a Scottish species in 1983 after some confusion of identity.

Heaths replace mires in better drained acidic conditions where it is too dry for active peat building, but the transition between mire and heath may be gradual. Heath vegetation is dominated by ericaceous shrubs such as heather (*Calluna vulgaris*), bell heather (*Erica cinerea*) and blaeberry (*Vaccinium myrtillus*). The orchid flora is rather restricted with similar species represented as in mires, but there is no suitable habitat for *Hammarbya paludosa*. *Dactylorhiza maculata* subsp. *ericetorum* and *Listera cordata* are the most likely to be encountered.

The distinction between mire and fen is based mainly on topography. The latter, unlike mires, occur on level or imperceptibly sloping ground where the water table is high and relatively stagnant. Fens are not a characteristic feature of the Scottish landscape but are quite well represented in the Borders where small but numerous examples occur. Fens with fairly high nutrient levels (rich fens) can have a similar orchid flora to calcareous flushes and marshes.

Conservation

It is a short step from consideration of orchid habitats in Scotland (and in Britain generally) to realisation of how threatened much of our flora is. The danger is not just from habitat destruction. It has to be acknowledged that in the past collection of these beautiful plants by botanists and others has brought some to the verge of extinction. Perhaps the prime example is *Cypripedium calceolus* which, since its first mention in 1629 as a native British plant in John Parkinson's *Paradisi in Sole Paradisus Terrestris*, has been ruthlessly collected. It is now confined to a single British locality (in north England) while most of the many specimens collected for gardens have perished.

Changes in land-use have, however, probably been responsible more than anything else for the decline of many of our native orchids and other wild flowers. For example, many orchids grow in calcareous grassland, a habitat that has been severely reduced by agricultural intensification and conversion into arable land. In England the first major losses came between the second half of the eighteenth and the first half of the nineteenth century when the previously lightly cultivated downland became enclosed and large areas of rough grazing were ploughed. Since then chemical fertilisers, herbicides and reseeding of meadows and pastures have transformed vast areas of semi-natural grassland into highly productive, but species-poor, grassland leys. It is estimated that about 80% of calcareous grassland and 95% of neutral grassland has been significantly damaged since 1940.

Calcareous grassland is not so widespread in Scotland as in England but one of the main areas of occurrence is the machair of the Western Isles. The flora of the machair is a product of low intensity, traditional crofting agriculture, and to maintain these species-rich swards, great care is needed especially with such modern agricultural practices as silage-making. Effective communication between crofters and conservationists is important to minimise the chances of damage. In some circumstances confidentiality regarding rare orchids is justified, but if the users of the land are not aware of what grows there, disaster can result.

Dactylorhiza purpurella subsp. *purpurella* growing in a wet area of young coniferous woodland will eventually disappear as the trees develop and the site becomes drier and shaded. Roxburghshire, 17 vi 1989.

The survival of *Dactylorhiza majalis* subsp. *scotica* depends on the continuation of traditional farming methods. North Uist, 8 vi 1991.

Grassland is only one habitat and many others, such as mire and moorland, have suffered through changes in land-use. Large tracts have been drained, ploughed and afforested. In Scotland *Hammarbya paludosa* tends to grow in botanically unremarkable mire and is very easily overlooked. This habitat is often difficult to conserve on botanical grounds and over large areas has been afforested. Scotland is a stronghold for the bog orchid and it is classed as vulnerable in Europe (by the International Union for the Conservation of Nature and Natural Resources). It could become seriously endangered if destruction or disturbance of mire continues.

Legislation has started to play an important role in protecting individual plants and preserving sensitive sites from development. The Wildlife and Countryside Act 1981 contains a schedule of plants that are legally protected from picking, uprooting or destruction. Currently there are over 90 scheduled plants of which ten are orchids. The Act also forbids the disturbance of any plant without the permission of the landowner. How effective this legislation continues to be depends to a considerable extent on the understanding and sympathy of landowners and tenants. Effective protection measures, whether they are imposed by law or not, require knowledge of the biology of the plants at risk and of their interactions with the environment. Ecological research is therefore vital to the conservation of rare plants and, indeed, to the prevention of commoner species becoming rare.

Artificial propagation is another measure that can help preserve rare species. This technique is still in its infancy for British wild orchids but some small natural populations have recently been successfully restocked with artificially propagated plants.

Perhaps the most important element in halting the decline of our rare plants is education. Even professional botanists may need reminding that trampling and disturbance can jeopardise the survival of the objects of their study. We hope this book will help to increase awareness of the beauty, diversity and fragility of our Scottish orchid flora.

3. ORCHID
Recording

Orchid Recording

The present system of plant recording in the British Isles has been in place since 1852 and uses a map of Britain and Ireland divided into areas of more or less equal size. These areas are called vice-counties (v.c.s) and are numbered 1 to 71 for England and Wales, 72 to 112 for Scotland, 113 for the Channel Islands, and H1 to H40 for Eire and Northern Ireland. In the main, the boundaries of these vice-counties follow those of the old county councils though the larger counties are sub-divided. The distribution maps featured throughout this book use this system to illustrate past and present distribution patterns.

British plant records are held by the Biological Records Centre at Monks Wood Experimental Station which is funded by the Natural Environment Research Council and Joint Nature Conservation Committee. The Centre collects data from a wide range of amateur recorders, Botanical and Natural History societies, and herbaria. A major role in recording is played by the network of vice-county recorders organised by the Botanical Society of the British Isles (BSBI). The Society's quarterly journal *Watsonia* contains general details of new and rare plants including, of course, orchids. In Scotland the BSBI produces a newsletter and organises exhibitions and meetings for members and recorders.

Data on the distribution of any wild plant usually include cases where records have never been substantiated, or where plants once found in a particular locality have never been re-discovered. This may reflect extreme rarity, or extremely erratic flowering. Alternatively, the site may have changed to such an extent that the species has died out. Occasionally anomalous records have been proven to follow deliberate introductions or hoaxes. Often, however, such records have simply been based on misidentification. Orchids have not escaped these problems. Indeed they may well have suffered more than most other groups.

Orchis morio is a good example of a species that seems to have disappeared from a locality. Currently recorded from only one place in south-west Scotland, this species was discovered some years ago growing in central Banffshire. Although the original identification was confirmed, the plants have never been re-found. This species thrives in calcareous grassland, a habitat which is uncommon in central Banffshire but does occur in a few areas of the county. In recent years another rare orchid, *Gymnadenia conopsea* subsp. *conopsea*, has been found growing in calcareous grassland, in the same general area where *Orchis morio* had been previously found. Moreover, a third lime-loving orchid, *Epipactis atrorubens*, has also been recently recorded from approximately the same area. It seems quite possible, therefore, that *Orchis morio* does indeed survive in Banffshire and may yet be re-discovered there.

A more venerable record that may represent a plant that has completely vanished from the Scottish orchid flora is that of *Serapias grandiflora* mentioned by John Lightfoot in his *Flora Scotica*. He found a single plant growing in a wood on the Isle of Arran in 1772. His description closely matches the species currently known as *Cephalanthera damasonium* which, in Britain, is now found only in southeast England.

A more contentious case is the 'Rhum heath spotted-orchid', *Dactylorhiza maculata* subsp. *rhoumensis*, which is known only from the island of Rhum. It was

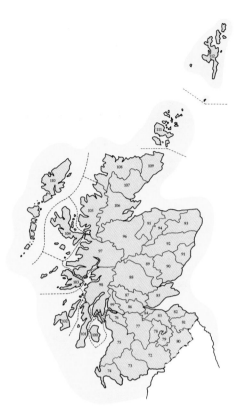

VICE-COUNTIES OF SCOTLAND

72 Dumfriesshire
73 Kirkcudbrightshire
74 Wigtownshire
75 Ayrshire
76 Renfrewshire
77 Lanarkshire
78 Peeblesshire
79 Selkirkshire
80 Roxburghshire
81 Berwickshire
82 East Lothian
83 Midlothian
84 West Lothian
85 Fifeshire
86 Stirlingshire
87 West Perthshire
88 Mid Perthshire
89 East Perthshire
90 Angus
91 Kincardineshire
92 South Aberdeenshire
93 North Aberdeenshire
94 Banffshire
95 Moray
96 East Inverness-shire
97 West Inverness-shire
98 Argyll Main
99 Dunbartonshire
100 Clyde Isles
101 Kintyre
102 South Ebudes
103 Mid Ebudes
104 North Ebudes
105 West Ross
106 East Ross
107 East Sutherland
108 West Sutherland
109 Caithness
110 Outer Hebrides
111 Orkney Islands
112 Shetland Islands

originally described in 1948 but many botanists consider that specimens do not differ sufficiently from the very variable subsp. *ericetorum* to merit either subspecific or even varietal status.

No discussion of anomalous Scottish records would be complete without mention of the calypso orchid, *Calypso bulbosa*, and the lady's slipper orchid, *Cypripedium calceolus*. The former is circumboreal, occurring in northern Scandinavia, north Russia, Asia and North America. Its supposed occurrence in the north of Scotland rests upon a reference in a French book, *Les Orchidées Rustiques*, written by Henry Correvon and published in 1893. Since Correvon gave no clue as to his source of information and there is no record of any herbarium specimen, it is impossible to validate this claim. It is certainly possible that a species with such a wide northern distribution might once have occurred in native Scots pine or birch woodlands. Nevertheless, that such a beautiful orchid could have escaped collection is difficult to believe and, had it been collected, specimens would surely have survived somewhere among our many excellent British herbaria.

A 'Scottish' specimen of *Cypripedium calceolus*, the lady's slipper orchid, does exist, collected near Cupar in Fife. However, the British Museum established that it was of a continental form and the likelihood is that it was planted as a hoax. In Britain the only current locality for this extremely rare orchid is a single site in northern England.

All apparently new records should be approached with caution as there are already several instances where records are based on misidentification of commoner plants. The study of Scottish orchids is never dull!

We hope that this book will stimulate interest in orchids and, if you find any that you think may not have been recorded, please contact the relevant vice-county recorder who will verify the record and if necessary add it to the list of known localities for the species. Please refrain from collecting any part of the plant. The names and addresses of all recorders can be obtained from the BSBI, c/o The British Museum (Natural History), London or from the Royal Botanic Garden Edinburgh.

4. *Classification* of SCOTTISH NATIVE ORCHIDS

Classification of Scottish Native Orchids

SUBFAMILY Orchidoideae
 TRIBE Neottieae
  SUBTRIBE Neottiinae
1. *Cephalanthera longifolia* (L.) Fritsch
2. *Epipactis palustris* (L.) Crantz
3. *Epipactis helleborine* (L.) Crantz
4. *Epipactis youngiana* A. J. Richards & A. F. Porter
5. *Epipactis leptochila* (Godfery) Godfery var. *dunensis* T.& T. A. Stephenson
6. *Epipactis atrorubens* (Hoffman) Besser
7. *Spiranthes romanzoffiana* Chamisso
8. *Listera ovata* (L.) R. Brown
9. *Listera cordata* (L.) R. Brown
10. *Neottia nidus-avis* (L.) L. C. M. Richard
11. *Goodyera repens* (L.) R. Brown

 TRIBE Epidendreae
  SUBTRIBE Liparidinae
12. *Hammarbya paludosa* (L.) O. Kuntze

 TRIBE Vandeae
  SUBTRIBE Corallorhizinae
13. *Corallorhiza trifida* Chatelin

 TRIBE Orchideae
14. *Coeloglossum viride* (L.) Hartman
15a. *Gymnadenia conopsea* (L.) R. Brown subsp. *conopsea*
15b. *Gymnadenia conopsea* subsp. *densiflora* (Wahlenberg) Camus,
        Bergon & A. Camus
15c. *Gymnadenia conopsea* subsp. *borealis* (Druce) F. Rose
16. *Pseudorchis albida* (L.) A. & D. Löve
17. *Platanthera chlorantha* (Custer) Reichenbach
18. *Platanthera bifolia* (L.) L. C. M. Richard
19. *Orchis morio* L.
20. *Orchis mascula* (L.) L.
21a. *Dactylorhiza fuchsii* (Druce) Soó subsp. *fuchsii*
21b. *Dactylorhiza fuchsii* subsp. *hebridensis* (Wilmott) Soó
21c. *Dactylorhiza fuchsii* subsp. *okellyi* (Druce) Soó
22. *Dactylorhiza maculata* (L.) Soó subsp. *ericetorum* (E. F. Linton)
        P. F. Hunt & Summerhayes
23a. *Dactylorhiza incarnata* (L.) Soó subsp. *incarnata*
23b. *Dactylorhiza incarnata* subsp. *coccinea* (Pugsley) Soó
23c. *Dactylorhiza incarnata* subsp. *pulchella* (Druce) Soó
23d. *Dactylorhiza incarnata* subsp. *cruenta* (O. F. Mueller) P. D. Sell
24. *Dactylorhiza majalis* (Reichenbach) P. F. Hunt & Summerhayes
        subsp. *scotica* E. Nelson
25a. *Dactylorhiza purpurella* (T. & T. A. Stephenson) Soó subsp. *purpurella*
25b. *Dactylorhiza purpurella* subsp. *majaliformis* E. Nelson
26. *Dactylorhiza lapponica* (Hartman) Soó
27. *Dactylorhiza traunsteineri* (Sauter ex Reichenbach) Soó
28. *Anacamptis pyramidalis* (L.) L. C. M. Richard

Opposite Platanthera bifolia
growing with *Pedicularis sylvatica*.
North Uist, 15 vi 1990.

5. *Field Key to* SCOTTISH NATIVE ORCHIDS

Field Key to Scottish Native Orchids

Patrick Woods and Mary Bates

When identifying an orchid, several points should be borne in mind. Flowers, although they may vary in minor details such as intensity of colour, are the most reliable feature. Plants in exposed situations are often smaller than those in, for example, sheltered positions or in tall vegetation. Pigmentation, particularly the reddish colour from anthocyanins in stems and ovaries, is affected by light levels, usually being stronger in brighter light. Scent can not only be a rather subjective character but can also be influenced by the age of the flower, whether pollination has taken place, the weather, and even the time of day. Dactylorchids are difficult to separate, particularly at subspecies level, and further complications arise because they hybridise freely. Here geographic and habitat differences may be useful. *Epipactis youngiana* and *E. leptochila* var. *dunensis* are very similar in appearance and are insufficiently known in Scotland to be identified with certainty.

We have followed the most widely accepted current nomenclature, but the correct application of names and the taxonomic rank of some of the dactylorchids is still a matter of debate among botanists.

This key and the species accounts are based entirely on Scottish plants, which differ in some minor respects from material from other parts of Britain.

Artist Mary Bates mixes paint to match plant colour before completing a painting of *Orchis morio*. Ayrshire, 17 v 1991.

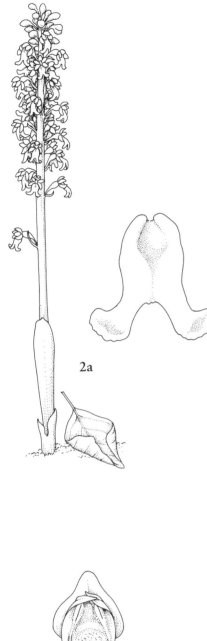

2a

1a. Plants without green leaves _____2
1b. Plants with green leaves _____3

2a. Plants robust, entirely creamy-brown
or straw-coloured, to 52cm tall;
inflorescence many-flowered, crowded;
lip c.12mm long, deeply cleft into
2 lobes, honey-brown, unspotted
                            10. Neottia nidus-avis
2b. Plants slender, yellowish or reddish, to 28cm
tall; inflorescence 4–13-flowered,
not crowded; lip c.5mm long, shallowly
3-lobed, white with crimson spotting
towards base _____*13. Corallorhiza trifida*

2b

3a. Lip with spur _____4
3b. Lip without spur _____26

4a. Spur distinctly longer than ovary,
slender _____5
4b. Spur as long as or shorter than ovary,
not slender _____10

5a. Lip strap-shaped, not lobed; flowers always
white or greenish white, sweetly scented; leaves
2, rarely more, sub-opposite, broad elliptic to
ovate _____6
5b. Lip lobed; flowers usually bright pink or reddish
purple, rarely white, pleasantly or unpleasantly
scented; leaves several, not sub-opposite,
lanceolate or oblong _____7

6a. Petals curving round to form a semi-circular
hood over column; pollinia divergent; entrance
to spur clearly seen
                            17. Platanthera chlorantha
6b. Petals forming a loose triangular hood over
column; pollinia parallel; entrance to spur not
clearly seen _____*18. Platanthera bifolia*

6a

6b

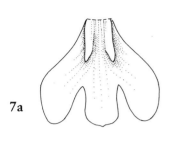

7a

7a. Inflorescence conical, less so when mature, with
slightly unpleasant foxy smell; lip deeply
3-lobed with 2 plate-like ridges at base; leaves
grey-green _____**28. *Anacamptis pyramidalis***

7b. Inflorescence cylindrical, sweetly scented; lip
with 3 rounded lobes, not ridged at base; leaves
shiny bright green _____8

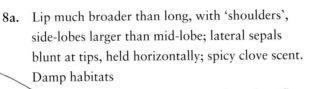

8a. Lip much broader than long, with 'shoulders',
side-lobes larger than mid-lobe; lateral sepals
blunt at tips, held horizontally; spicy clove scent.
Damp habitats
*15b. **Gymnadenia conopsea** subsp. **densiflora***

8b. Lip as long as or slightly longer than broad,
side-lobes equal to or smaller than mid-lobe;
lateral sepals pointed at tips, angled downwards;
scent slightly acidic or of carnations.
Drier habitats _____9

8a

9a

9a. Lip with 3 almost equal lobes; lateral sepals
linear; scent with acidic overtones. Dry limestone
areas
*15a. **Gymnadenia conopsea** subsp. **conopsea***

9b. Lip with side-lobes shorter than mid-lobe, lobes
sometimes hardly noticeable; lateral sepals oval-
lanceolate; carnation-scented. Hill pastures
*15c. **Gymnadenia conopsea** subsp. **borealis***

10a. Flowers creamy or greenish white or greenish, or
suffused brownish or purplish; lip with 3 apical
lobes _____11

10b. Flowers purplish, reddish or rarely white; lip not
as above _____12

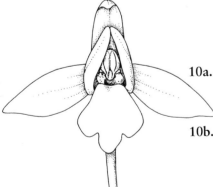

9b

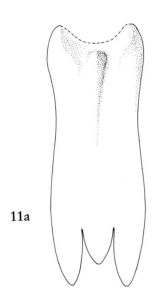

11a

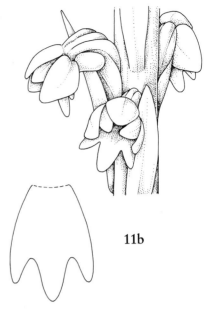

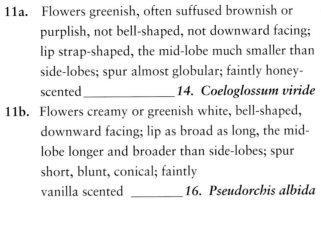

11a. Flowers greenish, often suffused brownish or purplish, not bell-shaped, not downward facing; lip strap-shaped, the mid-lobe much smaller than side-lobes; spur almost globular; faintly honey-scented _____ **14. *Coeloglossum viride***

11b. Flowers creamy or greenish white, bell-shaped, downward facing; lip as broad as long, the mid-lobe longer and broader than side-lobes; spur short, blunt, conical; faintly vanilla scented _____ **16. *Pseudorchis albida***

11b

12a. Basal leaves in a rosette, upper leaves clasping stem for almost their whole length; bracts somewhat membranous, 1–2 mm wide, often reddish _____ **13**

12b. Leaves not arranged as above, upper leaves not clasping stem; bracts ± leaf-like, 3–5mm wide _____ **14**

12a

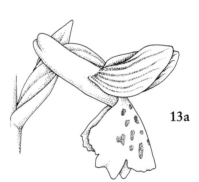

13a. Stem 6–15cm tall; inflorescence 3–12-flowered; lateral sepals angled forward, distinctly green-veined, sepals and petals forming a loose hood; lip broader than long, 3-lobed to almost entire, mid-lobe not longer than side-lobes; leaves not purple marked; usually sweetly scented _____ **19. *Orchis morio***

13a

13b. Stem 8–46cm tall; inflorescence many-flowered; lateral sepals angled upward, not distinctly veined, dorsal sepal and petals forming a loose hood; lip as broad as long, 3-lobed, mid-lobe longer than side-lobes; leaves of most plants with elongated purple blotches; often smells of tom-cats, occasionally sweetly scented **20. *Orchis mascula***

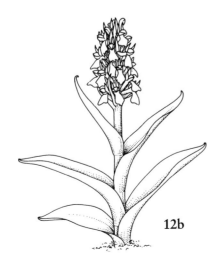

12b

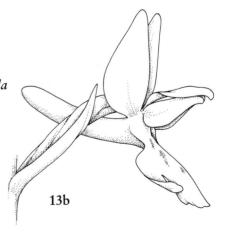

13b

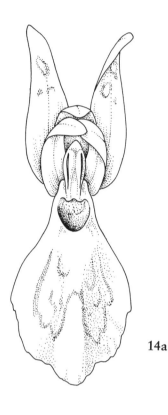

14a. Lateral sepals angled upwards, often touching back to back; sides of lip usually reflexed to some extent; spur over 2mm diameter; flowers flesh pink, dark red, magenta to purple; leaves purple marked or not; non-sheathing leaves few, usually 1, occasionally to 3; lower floral bract more than 3mm wide _____ 15

14b. Lateral sepals spreading or angled below the horizontal; side-lobes of lip not reflexed; spur less than 2mm in diameter; flowers predominantly whitish or pinkish, sometimes darker pink or reddish purple, or white; leaves usually purple marked; non-sheathing leaves to 3 or more; lower floral bract no more than 3 mm wide _____ 23

15a. Plants slender; lower sheathing leaves linear to lanceolate, tips not hooded; bracts purplish, purple-marked or not; inflorescence lax; flowers magenta-purple, magenta-red, magenta-pink or lilac _____ 16

15b. Plants usually robust; lower sheathing leaves lanceolate to ovate-lanceolate, tips hooded or not; bracts green, reddish or purplish, rarely purple-marked; inflorescence dense; flower colour as above or flesh colour or pink _____ 17

16a. Sheathing leaves 2–3, heavily blotched or spotted, purple-edged; non-sheathing leaves 0–2; bracts purple-marked on both sides; flowers magenta-purple or magenta-red, rarely lilac; lateral sepals erect and blunt with dark markings; lip markings strong and intense dark violet-purple _____ *26. Dactylorhiza lapponica*

16b. Sheathing leaves 2–5, almost always 3, usually unmarked, or if purple-marked then in upper ⅓ only; non-sheathing leaves 0–1; bracts unmarked; flowers magenta-purple to magenta-pink or lilac; lateral sepals upward spreading and pointed, faintly marked; lip marked with dots and lines, not intense nor tending to violet

   27. Dactylorhiza traunsteineri

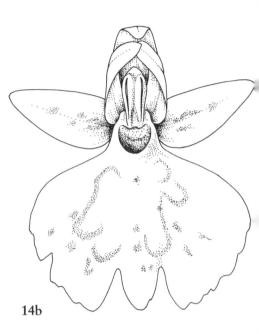

14b

16a

16b

14a

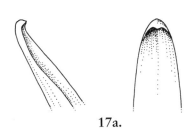

17a. Sheathing leaves ± erect, hooded at tips, usually yellowish green, usually without purple marks, if marked then on both sides of leaf; non-sheathing leaves 0–1; flowers flesh-pink, dark red, magenta to purple; lip unlobed or not deeply lobed, sides usually reflexed _____ **18**

17a.

17b. Sheathing leaves ± spreading, not hooded at tips, not yellowish green, blotched, spotted or unmarked, if marked then on upper surface only; non-sheathing leaves 1–2; flowers bright magenta-purple to pinkish purple; lip broad, flat, distinctly or indistinctly 3-lobed, mid-lobe small _____ **21**

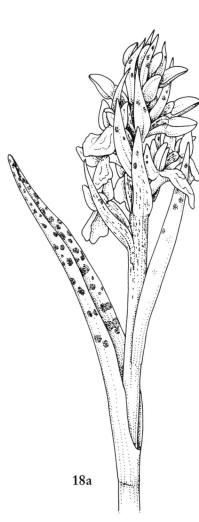

18a. Flowers pale lilac-purple to violet-magenta; lip not deeply lobed but mid-lobe usually prominent, side-lobes not strongly reflexed; leaves purple-spotted on both sides; fleck markings on upper stem, bracts and ovaries. NW Scotland

     *23d. **Dactylorhiza incarnata** subsp. **cruenta***

18b. Flowers flesh-pink, reddish purple, magenta to deep crimson; lip shallowly lobed to entire, mid-lobe not prominent, side-lobes strongly reflexed; leaves not purple-marked; upper stem, bracts and ovaries without flecks _____ **19**

19a.

18a

19a. Flowers flesh pink; plant 7–50cm tall, rarely more than 20cm. Marshy areas

    *23a. **Dactylorhiza incarnata** subsp. **incarnata***

19b. Flowers reddish purple, magenta to deep crimson; plant usually less than 20cm tall ___**20**

20a.   Flowers bright crimson to madder red; short
       stout plants 8–15cm tall; leaves thick, dark
       green, fairly broad, strongly keeled and hooded.
       Dune slacks and moist dune grassland, rarely a
       short distance inland
            23b. Dactylorhiza incarnata subsp. *coccinea*

20b.   Flowers mauve-purple or magenta;
       plants to 30cm tall; leaves pale green, narrow,
       not strongly keeled. Mainly in acid boggy areas
            23c. Dactylorhiza incarnata subsp. *pulchella*

21a.   Sheathing leaves 3, sometimes 4, crowded
       towards base, markings varying from smallish
       blotches to a complete purple-brown
       colouration, can be purple-edged; inflorescence
       not flat topped; flowers violet-purple;
       lip broad, distinctly 3-lobed. N Uist and
       possibly Jura
            24. Dactylorhiza majalis subsp. *scotica*

21b.   Sheathing leaves usually 4 or more, not crowded
       at base, unmarked or purple-marked;
       inflorescence flat-topped; flowers bright purple
       to pinkish purple or magenta; lip diamond-
       shaped or not deeply lobed. Widespread_____22

22a.   Leaves unmarked or occasionally with very small
       spots, usually towards apex; bracts unspotted;
       lip tending to be diamond-shaped; flowers
       bright purple to magenta. Widely distributed
            25a. Dactylorhiza purpurella subsp. *purpurella*

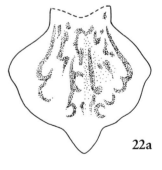

22a

22b.   Leaves usually with round purple blotches, often
       large, and usually over entire upper surface;
       bracts spotted or very rarely unspotted; lip
       tending to be 3-lobed; flowers pinkish purple.
       Outer Hebrides,
       N and NW coasts
            25b. Dactylorhiza purpurella subsp. *majaliformis*

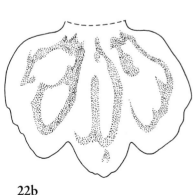

22b

23a

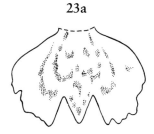

23a. Plants slender; leaves lanceolate, usually with round pale purplish spots; lowest leaf shorter but not wider than others, tip pointed; inflorescence few-flowered, ± pyramidal; flowers white to pinkish lilac, normally pale; mid-lobe of lip usually shorter than conspicuous side-lobes; spur 1mm wide. Acid soils and heathland

      22. *Dactylorhiza maculata* subsp. *ericetorum*

23b. Plant usually but not always robust; leaves narrow to broad, usually with dark transversely elongated blotches, or unmarked, lowest leaf shorter and wider than the others, tip rounded; inflorescence few- to many-flowered, short to long and tapering; flowers pale lilac to rosy purple, rarely white; mid-lobe of lip longer than side-lobes; spur 1–2mm wide. Mostly on alkaline soils, sometimes open woodland ____24

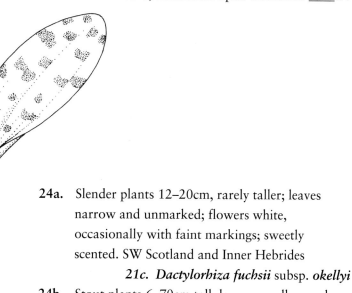

23a

23b

24a. Slender plants 12–20cm, rarely taller; leaves narrow and unmarked; flowers white, occasionally with faint markings; sweetly scented. SW Scotland and Inner Hebrides

      21c. *Dactylorhiza fuchsii* subsp. *okellyi*

24b. Stout plants 6–70cm tall; leaves usually purple-marked; flowers palest pink to rosy purple (very occasionally ± white); scarcely scented _____25

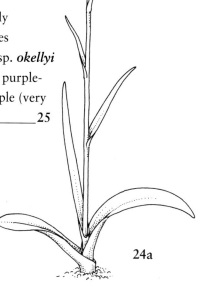

24a

25a

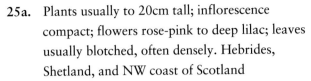

25a. Plants usually to 20cm tall; inflorescence compact; flowers rose-pink to deep lilac; leaves usually blotched, often densely. Hebrides, Shetland, and NW coast of Scotland
      21b. *Dactylorhiza fuchsii* subsp. *hebridensis*

25b. Plant 17–70cm tall; inflorescence elongated, cylindrical to tapering; flowers pale to deep pinkish purple (rarely white); leaves usually marked with transversely elongated blotches. Widely distributed
      21a. *Dactylorhiza fuchsii* subsp. *fuchsii*

26a. Leaves arising below middle of stem; upper part of stem bare _____27

26b. Leaves spaced along length of stem _____29

27a. Plant very small and slender, to 10cm tall; leaves to 1cm long, arising at or near swollen stem base; flower minute, yellowish green; lip uppermost, triangular, unlobed. Wet sphagnum bogs _____**12.** *Hammarbya paludosa*

27b. Leaves 2 (rarely 3), sub-opposite, arising in lower half of stem but not at the unswollen base; lip lowermost, strap-shaped, deeply cleft into 2 lobes, sometimes also with 2 small side-lobes near base _____28

28a. Robust plant, 10–60cm or more tall; leaves 5–20cm long, oval; inflorescence of numerous greenish flowers; lip reflexed, lobes rounded at tips _____**8.** *Listera ovata*

28b. Slender plant to 24cm tall; leaves 1–2.5cm long, heart-shaped or oval; inflorescence of 3–15 reddish green flowers; lip not reflexed, lobes pointed at tips _____**9.** *Listera cordata*

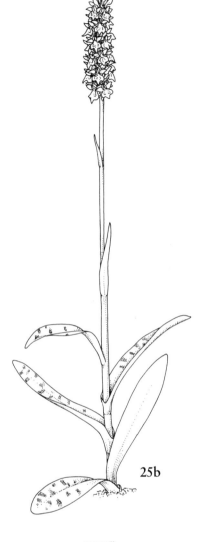

25b

27a

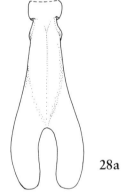

28a

28b

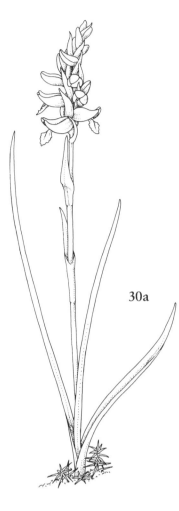

30a

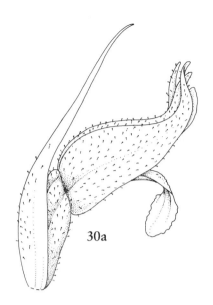

30a

29a. Flowers crowded, ± spirally arranged, glandular hairy, white or creamy-white; lip not divided into 2 parts by a joint or hinge _____ **30**

29b. Flowers not crowded, sometimes with downy hairs, never entirely white; lip divided into 2 parts: the apical part or epichile separated from the basal part or hypochile by a narrow fold or hinge _____ **31**

30a. Plants not creeping; leaves long, narrow and erect; inflorescence stout, 2–4 x 2cm; flowers in 3 spiral rows; apical edges of lip frilled and toothed; strongly hawthorn-scented. Mainly in damp places in W Scotland and the Hebrides
7. Spiranthes romanzoffiana

30b. Plants creeping, ± mat forming; leaves in a loose basal rosette and some way up stem, evergreen, stalked, ovate, usually conspicuously net-veined; inflorescence slender, 5–7 x 1cm; flowers in a single spiral row, twisted so that they tend to face the same direction; lip pointed at tip, sac-like at base; sweetly scented. Pinewoods in N and E Scotland _____ *11. Goodyera repens*

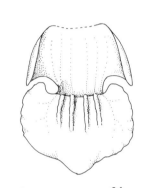

31a. Loose inflorescence of 3–40 white flowers tending to face upwards or horizontally; bracts, except sometimes lower one or two, tiny; sepals and petals not widely spreading; base of epichile with several basal orange ridges; ovary unstalked; leaves lanceolate to linear lanceolate
1. Cephalanthera longifolia

31b. Loose inflorescence of 7–60 greenish, purplish or brownish flowers, tending to face horizontally or downwards; bracts not tiny; sepals and petals widely spreading or not; base of epichile with boss-like swellings; ovary stalked; lower leaves ovate to elliptic _____ **32**

31a

30b

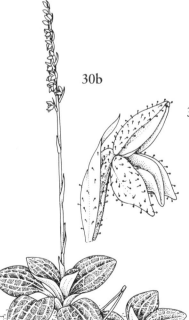

32a. Inflorescence of 7–14 showy flowers; sepals with
short hairs outside, brownish or purplish green,
paler inside; petals whitish suffused pink or
purplish; lip white, as long as or longer than
sepals and petals; epichile with frilled upturned
margins and yellow-edged boss at base,
connected by a narrow hinge to lobed, reddish
veined hypochile. Marshy places

2. Epipactis palustris

32b. Inflorescence of 3–60 not very showy flowers,
purplish, greenish or greenish suffused purplish
or pinkish; lip shorter than sepals and petals;
epichile without frilled margins, connected to
hypochile by a fold. Dryish open places or in
woodland _____ 33

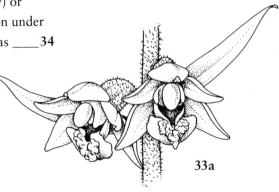

32a

33a. Flowers wine-red or purplish, very rarely cream;
epichile with reflexed tip and 3 prominent basal
bosses, hypochile green with red margin; upper
stem and ovary densely downy; lower leaves
usually suffused purplish on underside.
Restricted to limestone areas

6. Epipactis atrorubens

33b. Flowers not wine-red or cream, if purplish
always suffused green, sometimes pinkish or
whitish green; epichile with 2 or 3 basal bosses;
upper stem and ovary hairy (not downy) or
glabrous; leaves not suffused purplish on under
surface. Not restricted to limestone areas ____ 34

33a

34a. Plants usually large and robust; inflorescence of up to 60 drooping flowers, greenish flushed purplish or pink; epichile broader than long with reflexed tip and 2 rough basal bosses, hypochile dark brownish inside; pollinia remaining intact unless removed by an insect; leaves dark green, strongly ribbed underneath, lowest one broader than long _____ *3. Epipactis helleborine*

34b. Plants usually slender; inflorescence of up to 20 greenish, yellowish green or pink-tinged flowers; pollinia disintegrating in situ; leaves yellowish green, weakly ribbed underneath, lower ones longer than broad _____ 35

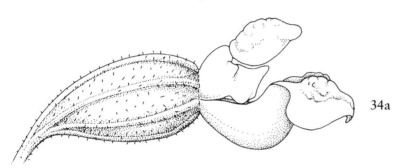

34a

35a. Sepals 8–11mm long, greenish, the margins usually faintly rose-coloured or white; petals greenish with rose margins; rostellum as long as anther; ovary sparsely short hairy or glabrous

4. Epipactis youngiana

35b. Sepals 6–8mm long, sepals and petals yellowish green; rostellum not more than half length of anther; ovary with short hairs

5. Epipactis leptochila var. *dunensis*

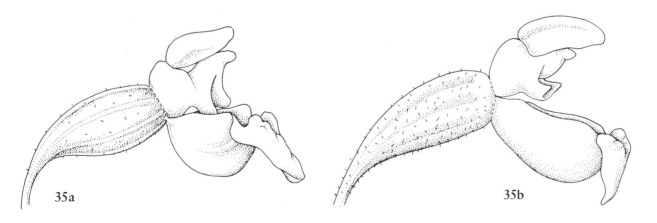

35a   35b

6. *The* ORCHIDS

Height: 15–60cm

Flowers:
Colour: pure white with a yellow patch at the base of the hypochile and 3–6 orange ridges on the epichile
Number: 3–40
Arrangement: moderately lax inflorescence; the lower flowers held horizontally away from the stem; lowermost bracts narrow, others tiny, much shorter than ovary
Sepals: long, pointed and same length as petals
Petals: as sepals
Lip: the ± heart-shaped hypochile is joined to the pointed, reflexed epichile by an elastic hinge
Ovary: thin, cylindrical and hairless

Leaves: c.7–20 lanceolate, arranged in a distinctive erect fan, upper leaves more linear, occasionally extending above flower spike

Stem: stout, a very few tiny glandular hairs near the top, otherwise hairless, with ridges on the upper section

Flowering: May to July

POST 1970

PRE 1970

1. *Cephalanthera longifolia* (L.) **Fritsch**

Synonyms: *Serapias helleborine* L. var. *longifolia* L.; *C. ensifolia* (Schmidt) L. C. M. Richard

Narrow-leaved Helleborine

The narrow-leaved helleborine is one of our most attractive orchids and is the only species of *Cephalanthera* present in Scotland. All other Scottish helleborines are species of *Epipactis*. When in full flower, the fine pure white flowers, with the lower ones held well away from the stem, are easily seen, even from a distance. Scottish plants normally have about 12 flowers but specimens can be found with up to 40. *C. longifolia* is restricted to a few areas in the west of Scotland, where it is normally found in open deciduous woodland or scrub.

Plants range from 15–60cm in height with several long, lanceolate, light green leaves arranged in a distinctive, erect fan. The tips of the uppermost leaves may extend above the inflorescences. Below the lower leaves of the fan are a few sheaths clasping the base of the rather stout stem; the upper leaves are more linear. The flowers often appear not to be fully open and the lower ones are held almost horizontally. Sepals and petals are all the same length with the lip divided into two parts, the basal hypochile and the apical epichile, which are connected by an elastic hinge. The hypochile has a small bright orange-yellow patch at the base and the epichile has 3–6 basal orange ridges. The flowers are mainly pollinated by small bees and are at their best during the last two weeks in May, although at some exposed localities some can still be found in the first week of July.

In Scotland *C. longifolia* is most commonly found growing in alkaline soil overlying limestone or on calcareous schist. It grows in a number of places from the Mull of Kintyre to Sutherland and Wester Ross, often under beech or mixed deciduous woodland or scrub. A number of populations have been recorded from the Inner Hebrides including Skye, Arran and Colonsay. Before 1970, the species was more widely distributed occurring in parts of central and eastern Scotland.

C. longifolia has a scattered distribution throughout the British Isles with records from Essex and Kent, east Wales, Cumbria, Durham and Ireland. Elsewhere in continental Europe it is found in most countries eastwards to Turkey and Greece. Further afield it has been recorded in North Africa, the Middle East and Japan.

No hybrids of *C. longifolia* have been recorded in Scotland since the only species that has ever been recorded as hybridising with it, the white helleborine (*C. damasonium*), does not now occur here.

Argyll, 31 v 1990

Argyll, 21 v 1991

Argyll, 5 vi 1991

Height: 15–45cm

Flowers:
Colour: purplish brown and white
Number: 7–14
Arrangement: loose, one-sided inflorescence, 7–15cm long; bracts lanceolate and pointed, lowest equalling the flowers, the rest shorter.
Sepals: ovate-lanceolate, pale green tinged purplish brown and hairy outside, paler inside
Petals: narrower and shorter than sepals, whitish tinged pink or purplish
Lip: predominantly white; hypochile red-veined with two triangular lateral lobes; epichile with wavy margin and yellow-edged boss at base
Ovary: narrowly pear-shaped, slightly downy

Leaves: 4–8, 5–15cm long, oblong-lanceolate, folded, often purplish below; uppermost narrow, pointed and bract-like

Stem: slightly downy in upper part

Flowering: June to August

2. *Epipactis palustris* (L.) Crantz

Synonyms: *Serapias helleborine* L. var. *palustris* L.; *S. palustris* (L.) Miller; *Helleborine palustris* (L.) Schrank

Marsh Helleborine

Epipactis palustris is one of Scotland's rarest native orchids. A lover of moist calcareous soils, it is mainly found in damp dune slacks where the presence of shell fragments help keep the pH level above 7. However, in 1983, a colony was discovered in the Central Highlands growing in damp upland grassy meadows overlying metamorphosed limestone rock.

Epipactis palustris is of medium height: Scottish specimens rarely exceed 20cm although occasional plants elsewhere in the British Isles reach 60cm. The upper part of the stem is slightly downy, and there are seven or eight oblong-lanceolate leaves with three to five prominent veins on the underside. In addition, the base of the stem often has one or two violet sheaths. Up to 14 large and showy flowers are carried in a loose, one-sided inflorescence. The three sepals are pale green tinged with purplish brown and the slightly shorter petals are dull white tinged with pink or purplish. The lip is mainly white and, as in all helleborines, is divided into two sections, a hypochile and an epichile. In this species the two are joined by a narrow elastic hinge. The hypochile is slightly concave with erect, triangular, red striped, lateral lobes, positioned at each side whereas the epichile is broader with a wavy margin and yellow-edged boss at its base. The narrow pear-shaped ovary is slightly downy. The marsh helleborine can be found in flower from late June until early August.

Epipactis palustris grows in damp calcareous meadows, dune slacks or, rarely, in disused gravel quarries. It is a rarity in Scotland, currently recorded from only two vice-counties, although before 1970 there were several more known localities, mainly in the south east and the Borders. It is more frequent in England, Wales and Ireland and occurs throughout continental Europe and Asia.

No hybrids of *E. palustris* have been recorded in Scotland, but it is known to have hybridised elsewhere in Europe with *Epipactis atrorubens* and *E. helleborine*, two species which can also be found in Scotland.

POST 1970

PRE 1970

Perthshire, 16 vii 1990

Perthshire, 16 vii 1990

Perthshire, 16 vii 1990

Height: 20–80cm

Flowers:
Colour: greenish to dull purple
Number: 10–60 rarely more
Arrangement: one-sided
inflorescence; bracts narrow,
pointed, lowest equal in length
to the flowers, uppermost
shorter equalling the ovary
Sepals: c.1cm, greenish purple,
ovate-lanceolate
Petals: somewhat narrower,
shorter and more pink than
sepals
Lip: hypochile cup-shaped, dull
reddish brown within; epichile ±
heart-shaped with reflexed tip
and two basal bosses; ranging
from purplish through pink to
greenish white
Ovary: pear-shaped, usually
± pubescent

Leaves: 4–10, ovate-elliptical,
dull green, becoming narrower
and pointed up stem, with c.5
strong ribs beneath

Stem: normally stout, with
short whitish hairs above and
flushed violet towards base.

Flowering: July to September

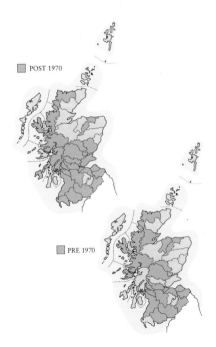

POST 1970

PRE 1970

3. *Epipactis helleborine* (L.) Crantz

Synonyms: *Serapias helleborine* L.; *Epipactis latifolia* (L.) Allioni

Broad-leaved Helleborine

The broad-leaved helleborine, though recorded from around 24 vice-counties in Scotland, must still be classed as uncommon. In recent years however, a number of new sites have been recorded within city boundaries. In Glasgow, for example, *E. helleborine* has been recorded from 40 of the 90 4km^2 tetrads which make up the area. Some botanists believe that *E. helleborine* may be commoner in Glasgow than anywhere else in Britain, perhaps because the increasing popularity of wildlife gardening has led to many gardens being allowed to revert to a semi-natural state. It is worth noting that, in Angus, the first record of the broad-leaved helleborine for more than 100 years was of a thriving clump on the edge of a somewhat neglected lawn in a garden in the centre of Brechin.

Epipactis helleborine is extremely variable, ranging in height from 20 to 80cm. Each of the one to three stems has between four and ten broad, ovate-elliptical leaves carried in a spiral arrangement, with two or three basal sheaths below. The flowers are borne in a one-sided inflorescence and can number up to 100 on more robust plants. Sepals and petals are quite broad and blunt-ended and range from green, the most usual colour, through pink to dull purple. The lip has a cup-shaped hypochile with a dull reddish brown inner surface, and a heart-shaped purplish, pinkish or greenish white epichile with a reflexed tip. Near the base of the epichile are two rough bosses which can aid identification when the flowers are fresh. The ovary is pear-shaped, usually ± pubescent but occasionally glabrous. *E. helleborine* can be found in flower from early July to mid-September.

The broad-leaved helleborine can thrive in an astonishing variety of habitats: dense woodland, shaded road verges, gardens, railway embankments, pit-bings, limestone outcrops and, in a few cases, among coastal dune slacks.

It is by far the most common of Scotland's helleborines, being found in most of the vice-counties of south and central Scotland. On the west coast it stretches as far north as Sutherland and occurs on some of the Inner Hebridean islands. In England and Wales it is even more frequent, occurring in all vice-counties except the Isle of Man. The Irish population is smaller and more scattered.

Further afield *E. helleborine* is found in most European countries and is widely distributed throughout Asia. In North America it is recorded, as an introduced species, from Quebec and Ontario and most of the eastern seaboard of the United States of America. Known as bastard helleborine, the plant was believed to have been introduced more than a century ago from Britain as a rather doubtful remedy for gout.

A hybrid with *E. atrorubens* has been reported from Sutherland. The other species with which hybrids might occur, are *E. leptochila* var. *dunensis* and *E. youngiana*, as all three species have recently been found growing in close association.

Perthshire, 4 viii 1990

Lanarkshire, 4 viii 1990

Perthshire, 4 viii 1990

Height: 20–58cm

Flowers:
Colour: green and rose pink
Number: 3–12 rarely more
Arrangement: one-sided inflorescence; lower bracts ± broad, upper bracts narrower
Sepals: ovate-lanceolate, greenish with rose-coloured or white margins
Petals: more ovate than sepals, greenish with rose-coloured margins
Lip: hypochile almost semi-circular, purple spotted inside; epichile heart-shaped with reflexed tip, rose with green central zone and two purple basal bosses
Stigma: tricornute with the central point, the rostellum, as long as the anther
Ovary: ovoid, ± hairless

Leaves: 4–7, pale yellowish green, lanceolate or ovate-lanceolate with wavy margins

Stem: slender but robust, hairy in upper part

Flowering: July

Mycorrhizal associate: material examined in 1992 showed the presence of typical orchid mycorrhiza but identification was not possible

POST 1970

4. *Epipactis youngiana* A. J. Richards & A. F. Porter

Young's Helleborine

This newly discovered species has a fascinating distribution on spoil-heaps in north England and Scotland. It was initially believed to be a hybrid between *Epipactis leptochila* and *E. helleborine*, and all three species were found together on wooded pit-bings in Lanarkshire in 1985. However, the original description of the species, in 1982, was based on finds in Northumberland where one of the supposed parents, *E. leptochila*, is not known. It is now suspected that the English *E. youngiana* may be a hybrid between *E. helleborine* and *E. phyllanthes* var. *pendula* which do grow together in other English and Welsh localities where *E. youngiana* has been identified. This would leave the possibility that the Scottish plant known as *E. youngiana* has a different origin from the English plant known as *E. youngiana*.

E. youngiana has solitary, erect stems 30–60cm tall. They are slender, pale yellowish green and smooth below the leaves but with short hairs above. The four to seven yellowish green leaves tend to be arranged in two opposing rows up the stem, and are usually narrow with wavy margins and without prominent veins beneath. There are also one or two basal sheaths. Up to 12 large, open, bell-shaped flowers are held in a one-sided inflorescence. The narrow ovate sepals are greenish with faintly rose-coloured or white margins. They are slightly longer than the petals which are more ovate and are greenish with rose margins. The small lip has an almost semi-circular hypochile, spotted purple inside. The rose-pink and green, heart-shaped epichile has a strongly reflexed tip and two purple basal bosses. The viscidium on the rostellum disappears as soon as the flowers open or even before, enabling the crumbling pollinia to fall directly onto the stigma. The ovoid ovary has only a few stout, prickly hairs. *E. youngiana* flowers during July.

The few locations so far recorded in Scotland indicate that *E. youngiana* favours open deciduous woodland on old pit-bings or spoil-heaps where the dominant tree cover is oak, ash, and birch, with a dense understorey of shrubs and regenerating trees. Elsewhere this species is known only from zinc and lead spoil-heaps in south Northumberland and also occurs in north and east Yorkshire and Glamorgan.

In Scotland a possible hybrid with *E. leptochila* var. *dunensis* has been noted and another with *E. helleborine* (which is present with *E. youngiana* at the Scottish sites) is known from Northumberland.

Lanarkshire, 17 i 1993

Lanarkshire, 3 viii 1990   Lanarkshire, 3 viii 1990

Height: 20–35cm

Flowers:
Colour: yellow-green
Number: 7–20
Arrangement: lax, one-sided inflorescence; bracts slender, shorter than flowers
Sepals: short and blunt, yellowish green without any pink flush
Petals: slightly shorter than the sepals, pink flushed
Lip: hypochile ± circular, mottled red-purple inside; epichile broadly triangular, as broad as long with a reflexed tip, whitish with a green or rosy tinge and two or three smooth basal bosses
Ovary: pear-shaped, ± hairy

Leaves: yellowish green, in two ± opposite rows, those towards the middle of the stem longest and broadest

Stem: slender, green, and hairy in upper part, violet-tinged and hairless towards base

Flowering: July

POST 1970

5. *Epipactis leptochila* (Godfery) Godfery var. *dunensis* T. & T. A. Stephenson

Synonym: *Epipactis dunensis* (T. & T. A. Stephenson) Godfery

Dune Helleborine

Although rare in Scotland, *E. leptochila* var. *dunensis* has been known for many years under the synonym *E. dunensis* from a few localities elsewhere in Britain. The Scottish populations have been found only at inland sites, whereas the English plants were first found in duneland habitats, and only more recently inland.

The slender solitary hairy stems reach 20–35cm in height. Seven to ten broadly lanceolate, pale yellowish green leaves are arranged in two more or less opposite rows up the stem; the bracts are slender and shorter than the flowers. Two or three basal sheaths are also present. The lax, one-sided inflorescence has 7–20 small, yellow-green flowers which characteristically droop when fully open. Petals and sepals are usually short and blunt; the sepals are almost always yellowish green, but the petals have a pink flush. The lip is pinkish green, with an almost circular hypochile mottled red-purple inside. The reflexed epichile is as broad as long and whitish with a green or rosy tinge. There are also two or three smooth basal bosses. In this species the rostellum lacks a viscidium, and is shorter than the anther. As the pollinia crumble almost as soon as the flowers open, self-pollination is usual. By contrast to *E. youngiana* the pear-shaped ovary is more or less hairy, and this provides a good character to separate the two species. Like *E. youngiana*, *E. leptochila* var. *dunensis* flowers in July.

In Scotland, as with *E. youngiana*, this species has only been found on old pit-bings or spoil-heaps south of Glasgow, with regenerating oak, ash and birch. Elsewhere in Britain the dune helleborine has also been found on inland zinc and lead spoil-heaps in North Durham, North Yorkshire and Lincolnshire. However, the typical, and by far the most widespread habitat, is moist hollows in open duneland. It is known in coastal dunelands in Anglesey, Lancashire and Northumberland (Lindisfarne). It has not been recorded from Ireland.

A plant found at one of the Scottish localities has been tentatively identified as a possible hybrid with *E. youngiana*.

This spoil-heap has remained undisturbed for almost a century. Lanarkshire, 17 i 1993

Lanarkshire, 3 viii 1990

Lanarkshire, 3 viii 1990

Height: 15–30(–60)cm

Flowers:

Colour: deep wine-red, very rarely creamy white

Number: 6–18(–28)

Arrangement: small flowers in a ± one-sided inflorescence; bracts narrow and pointed, lower equalling or longer than the flowers

Sepals: ovate, pointed, wine-red, greenish inside

Petals: ovate, blunt, wine-red to violet, incurved with the sepals

Lip: shorter than sepals and petals; hypochile green, red-spotted inside with red margin; epichile deep red or reddish violet with a reflexed tip and three bright red basal bosses

Ovary: pear-shaped, wine-red to violet, downy

Leaves: 5–10 in two ± opposite rows, ovate to elliptical, pointed, becoming narrower further up stem; all folded and rough to the touch; lower leaves usually suffused purplish on underside

Stem: solitary, upper part densely hairy, tinged violet towards base

Flowering: June to July

POST 1970

PRE 1970

6. *Epipactis atrorubens* (Hoffman) Besser

Synonyms: *Serapias atrorubens* Hoffman; *E. atropurpurea* Rafinesque; *E. rubiginosa* (Crantz) Gaudin

Dark-red Helleborine

The striking wine-red flowers of the dark-red helleborine are a wonderful sight when seen against limestone on cliff ledges or in the crevices of limestone pavement. However, since populations are subjected to intense grazing by sheep and deer only the flowering stems of plants growing in inaccessible places survive long enough to flower and set seed.

The normally single stem is from 15 to 30cm (rarely to 60cm) tall, softly hairy, densely so in upper part, violet tinged towards the base, and with one to three basal sheaths. The five to ten leaves are in two more or less opposite rows and are ovate elliptical and pointed, becoming narrower further up the stem. All the leaves are folded, many-veined, and feel rough on both sides. The lower leaves are usually suffused with purple on the underside. The bracts are lanceolate, the lower ones equalling or longer than the flowers. Each inflorescence carries 6–18 wine-red or violet-purple, rarely creamy white flowers which are faintly fragrant. The pointed ovate sepals are wine-red tinged with green whereas the petals are blunter and wine-red to violet. Both petals and sepals spread and curve inwards. The lip is shorter than the other perianth parts; the hypochile is green with a red margin and spots. The deep wine-red to violet epichile has a small, pointed, reflexed tip and three bright red, strongly wrinkled basal bosses. The ovary is 6–7mm long and pear-shaped. It is wine-red to violet, downy and droops when ripe. *E. atrorubens* flowers from June to July.

The distribution of *Epipactis atrorubens* follows that of exposed limestone in Britain where it grows in cliff crevices, on open stony screes or expanses of limestone pavement. In Scotland it is found in Wester Ross, Sutherland, Skye and Banffshire as well as East Perthshire. Elsewhere in Britain it is recorded in Yorkshire, the southern Lake District and North Wales. It is also found in County Clare and County Galway and is widespread in continental Europe.

A hybrid between *E. atrorubens* and *E. helleborine* has been recorded in Scotland. Although plants exposed to direct sunlight occasionally appear to be a washed-out colour, a true variant with cream-coloured flowers was recorded in 1960 from the Kishorn limestone area of West Ross. Today at the same locality many of the normally coloured plants have flowers with pointed sepals and longer than normal petals which give the flowers a more open appearance.

Skye, 18 viii 1990

Sutherland, 27 vii 1991

Sutherland, 27 vii 1991

Height: 10–30cm

Flowers:
Colour: creamy white
Number: 15–20 (–35)
Arrangement: dense inflorescence of large flowers in three spiral rows, but occasionally few-flowered, not dense and spirals indistinct; bracts narrow, pointed, with a few glandular hairs on the outer surface, lowermost as long as flowers
Sepals: lanceolate, creamy white flushed green at the base, with glandular hairs on the outer surface; forming a loose tube with the petals
Petals: narrow, blunt, forming loose tube with sepals
Lip: fiddle-shaped, with two minute basal bosses, apical edges frilled and toothed; white with green veins
Scent: strongly hawthorn-scented
Ovary: cylindrical, ± stalkless with scattered glandular hairs

Leaves: lower leaves: long, narrow, erect, yellowish green; upper leaves: short, pointed, lying close to stem, with loosely sheathing base

Stem: bluntly 3-angled, yellowish, upper part with scattered glandular hairs

Flowering: late July to early September

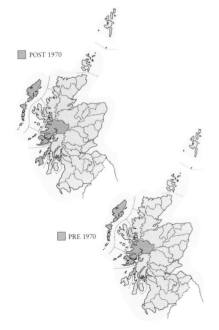

POST 1970

PRE 1970

7. *Spiranthes romanzoffiana* Chamisso

Mogairlean Bachlach Bàn
Irish Lady's-tresses

The small flowers arranged in tight spiral rows are said to resemble neatly braided hair, giving rise to the name lady's-tresses. *Spiranthes romanzoffiana* is one of the rarest orchids in Europe, being confined to small areas of Ireland and western Britain. The first Scottish record was made in 1921 on Coll in the Inner Hebrides (but under the name *S. autumnalis*), whereas the first record for Ireland was made in 1810. Irish lady's-tresses is now known from both island and mainland localities.

Plants of *S. romanzoffiana* found in Scotland vary in height from 10–30cm. They have up to eight narrow-lanceolate leaves which sometimes have inrolled edges, making them appear even narrower. The flowers are arranged in three spiral rows, and not in a single row as in other European species of *Spiranthes*. This arrangement of usually up to 20 large, creamy white, flowers gives the inflorescence a dense appearance. However inflorescences of Scottish plants may be few-flowered and less dense, with indistinct spirals. This species is usually found flowering in August but may appear as early as July or as late as September depending on weather conditions and water levels.

Irish lady's-tresses favours marshes, peaty pastures and flood plains which become inundated during winter or spring. The favourite habitats are purple moor-grass (*Molinia caerulea*) carpet or old lazy-beds subject to grazing. It is known from the islands of Coll, Colonsay, Islay and Mull, Barra, Benbecula and Vatersay, as well as the mainland areas of Ardnamurchan, Morvern, Moidart and Kintyre. *S. romanzoffiana* should be looked for in suitable habitats anywhere in western Scotland and the Islands. South of the border, it is only found in Devon and its main area of distribution is North America.

S. romanzoffiana is not known to hybridise in Britain or Ireland.

Coll, 22 viii 1991

Coll, 22 viii 1991

Inverness-shire, 9 viii 1989

Height: (10–) 20–60 (–75)cm

Flowers:
Colour: green, sometimes tinged reddish brown
Number: 15–30+
Arrangement: lax inflorescence of green flowers on short stalks; bracts short, about half as long as the flower stalks
Sepals: ovate, blunt, dull green sometimes tinged red, forming loose hood with petals
Petals: narrower, dull green, forming loose hood with sepals.
Lip: 7–15mm long, forked, lobes rounded at tips; bright yellow-green
Ovary: globular and ridged

Leaves: pair of subopposite large oval ribbed leaves, deep green to yellowish green; occasionally a smaller third leaf is present

Stem: glandular hairy above leaves, smooth below

Flowering: May to August

Mycorrhizal associate: unidentified species of *Ceratobasidium* isolated from population in Argyll

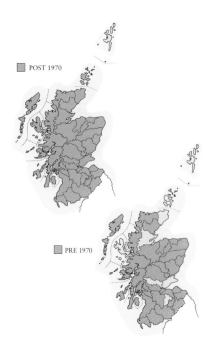

POST 1970

PRE 1970

8. *Listera ovata* (L.) R. Brown

Synonym: *Ophrys ovata* L.

Dà-dhuilleach
Common Twayblade

About twenty-five species of *Listera* occur in temperate North America and Asia but only two, *L. ovata* and *L. cordata*, are found in Europe. Common twayblade is one of the most widely distributed orchid species, and is easily identified by the two large, subopposite, oval leaves near the base of the stem, from which it derives its common name.

This species has a remarkably long life cycle, up to 15 years from germination to flowering, of which the first four years of establishment have to be in completely undisturbed soil. In these early years, seedlings are dependent on association with a mycorrhizal fungus. When plants reproduce vegetatively, as they are able to do from root buds, they flower within a few years. Even in well established colonies, the number of plants that appear above ground fluctuates widely. In part, this may be in response to soil moisture conditions.

Plants normally range in height from 20–60cm but occasional specimens may reach 75cm while smaller ones, down to 10cm or so can often be found in quite high numbers in the short turf of the Hebridean machair. The characteristic pair of oval leaves are prominently 3–5 ribbed and are normally between 7.5–10cm long by 5–7cm wide. Occasionally, an additional more pointed leaf may be found either above or below the main pair. A somewhat lax inflorescence, 7–25cm long, carries a large number of small flowers on short stalks. Extremely robust plants may have as many as 100 but between 15 and 30 is normal. The flowers are mostly dull green, sometimes tinged with reddish brown, and the sepals and petals form a loose hood. The lip is a much brighter yellow-green and larger than the other petals or sepals, being long, strap-like and forked to about half its length, occasionally with a small tooth between the lobes. The lip folds back underneath the flowers and there is a nectar-secreting groove running from the base of the lip to the fork. The ovary is globular and ridged. Pollination is by small insects and plants can be found in flower from May until late July or August.

Common twayblade can be found in an extremely wide variety of habitats but flourishes best in open, moist woodland on base-rich soils. In Scotland it thrives in open grassland and machair in the west, on dune slacks of the east coast, and the limestone pavement areas of the far northwest. It has been reported to compete successfully with bracken and heather on Orkney. Plants have also been found in 'new' habitats including old pit bings and disused railway lines, the latter seeming to provide ideal conditions for this pioneering species.

Listera ovata is recorded from every vice-county in Scotland except Shetland. In England and Wales it is even more abundant occurring in all vice-counties and extensively in moist woodlands, hill pastures and coastal dune slacks. It is also present throughout Ireland where again it is found in a great range of habitats. It is common throughout the rest of Europe occurring in almost all countries, and continues eastward into central Asia.

No hybrids have been recorded but plants with abnormal flowers are not uncommon. These variants may have all the petals shaped like the lip or conversely a lip shaped like the petals.

Lewis, 12 vii 1991

Roxburghshire, 26 vi 1990

Height: 5–24cm

Flowers:
Colour: reddish green
Number: 3–15
Arrangement: loose inflorescence of short-stalked tiny flowers, the lower ones occasionally in a whorl; bracts tiny, up to 1mm, triangular and greenish
Sepals: ovate, blunt, spreading and red-brown tinged green
Petals: narrower, also spreading, red inside, green outside
Lip: 3–4mm long, deeply forked, lobes pointed at tip, sometimes two tiny side-lobes near base; deep red-brown, projecting downwards
Ovary: spherical, green with reddish ribs

Leaves: two opposite heart-shaped leaves, ⅓–½ way up the stem; very rarely a third leaf present

Stem: slender, hairless below, slightly angular with glandular hairs above leaves

Flowering: May to July; withered flowers persisting until September

9. *Listera cordata* (L.) R. Brown

Synonym: *Ophrys cordata* L.

Dà-dhuilleach Monaidh
Lesser Twayblade

Listera cordata is very much the common twayblade, *Listera ovata*, in miniature. However, the leaves differ in being very much smaller and heart-shaped. Plants are normally only 5–24cm tall. The two opposite leaves are located about a third of the way up the stem which has fine glandular hairs for a short distance above them. The leaves are dark shiny green above and a lighter green beneath. It is often found under rank heather, where the leaves of non-flowering plants may be confused with those of blaeberry, *Vaccinium myrtillus*. The loose inflorescence has between three and 15 tiny reddish green flowers with the relatively large forked lip deep red-brown. Sepals and petals are spreading and, with the deeply divided lip, give the flower a star-like appearance. The ovary is fairly conspicuous, being spherical, ribbed and large in proportion to the rest of the flower. Pollination is by small insects and *L. cordata* also reproduces vegetatively by budding from the roots. This species can be found in flower from May until mid-July.

The most common habitat for *L. cordata* is heather moorland where, sheltered under heather in a dark micro-climate, the tiny orchid grows through a cushion of damp moss, with the only competition being the odd blaeberry or straggling tormentil (*Potentilla erecta*). It is always worthwhile to search in any area of rank heather for this species. It also grows among moss in damp, dark woodland, occasionally pinewoods, but more commonly willow / alder / birch carr.

Scotland is the main British stronghold for the lesser twayblade, where it has been recorded from all vice-counties with the exception of West Lothian. In many areas *L. cordata* can be considered locally abundant but, because of its size and the dark habitats where it is found, this tiny species can frequently be overlooked. Lesser twayblade can also be found in northern England, north Wales and the moorland areas of Devon and Somerset and occurs throughout Ireland, more frequently in the north. It occurs in all Scandinavian countries and at altitudes up to 2000m in the Alps and mountainous areas of Yugoslavia, Greece and Turkey. The species has a circumboreal distribution in northern Asia, Greenland and North America.

No hybrids have been recorded for this species.

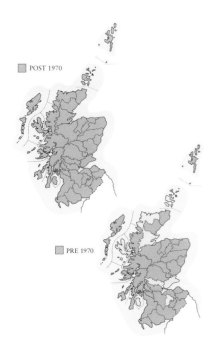

POST 1970

PRE 1970

Listera cordata growing in willow/alder/birch carr.
Angus, 30 vi 1989

Inverness-shire, 16 vi 1990

Inverness-shire, 16 vi 1990

Height: 17–52cm

Flowers:
Colour: creamy brown
Number: 20–80+
Arrangement: dense inflorescence, 7–15cm long, more open below; bracts pointed, papery and slightly shorter than the ovaries
Sepals: ovate-oblong, brown, forming loose hood with the petals
Petals: slightly shorter, brown
Lip: c.12mm long, slightly darker brown, deeply cleft; the lobes spreading and rounded at the ends
Ovary: ovoid, hairless
Scent: sickly honey fragrance

Leaves: none; a few large light brown sheathing scales

Stem: slightly glandular hairy

Flowering: May to July

10. *Neottia nidus-avis* (L.) L. C. M. Richard

Synonym: *Ophrys nidus-avis* L.

Bird's-nest Orchid

There are a number of species of *Neottia* scattered throughout Asia but *N. nidus-avis* is the only species known in Europe. This saprophytic plant is extremely well camouflaged, the light brown plants blending in with the surrounding leaf litter. Indeed, when searching for this uncommon species it is often easier to spot the darker brown dead inflorescences of the previous year. *N. nidus-avis* relies almost entirely for food on a mycorrhizal association.

Flowering plants as tall as 52cm have been reported but normally plants are between 20 and 40cm high. They are leafless although the slightly downy stem has a number of large light brown sheathing scales. Flowers at the top of the compact inflorescence are more densely grouped than the lower ones, and there is often a single flower quite far down the stem. Individual flowers are creamy brown; the sepals and petals form a loose hood, and the slightly darker lip is deeply divided into two spreading lobes with rounded tips. At the base of the lip is a small depression which secretes nectar. The bright yellow pollinia stand out in contrast to the soft brown flowers which have a scent reminiscent of honey. Pollination is thought to be by small flies and other insects but self-pollination does occur and the plants are said to reproduce vegetatively from buds at the root tips. *N. nidus-avis* flowers from May to July.

In Scotland the main habitat for *N. nidus-avis* is in beech woods, but occasionally plants may be found under mixed deciduous trees or, rarely, under yew.

The bird's-nest orchid is not common in Scotland. However, it can be found in around 60% of our vice-counties though there is no clear pattern of distribution. It is present on Mull but not known from any of the other Hebridean Islands, or Orkney or Shetland. Slightly more common south of the border, it is best known in the beechwoods of the southeast. In Ireland and Wales distribution is decidedly local. Distribution elsewhere in Europe is widespread but local in most central and southern countries. The species spreads eastwards as far as central and northern Asia, and is also recorded from Japan.

No hybrids have been recorded from Scotland.

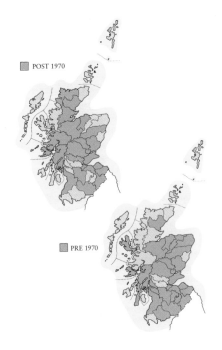

POST 1970

PRE 1970

Argyll, 31 v 1990

Neottia nidus-avis with young plants of *Cephalanthera longifolia*.   Argyll, 31 v 1990.

Height: 8–35cm

Flowers:
Colour: creamy white
Number: 5–25
Inflorescence: loose single
spiral; bracts narrow, 10–15mm
long, exceeding the length of the
ovary, with scattered glandular
hairs towards base on outside
Sepals: ovate and blunt, covered
on outside with glandular hairs;
the dorsal forming a tight hood
with the petals; may be tinged
with green
Petals: narrower, forming tight
hood with dorsal sepal
Lip: shorter than the sepals; sac-
like at base, pointed at tip
Ovary: almost stalkless,
c.10mm long, glandular hairy
Scent: sweetly fragrant.

Leaves: 3–5 forming a basal
rosette; stalked, ovate, dark
green with paler net-veining

Stem: glandular hairy in upper
part

Flowering: July to August

Mycorrhizal associate:
Rhizoctonia goodyera-repentis
isolated from populations in
Aberdeenshire

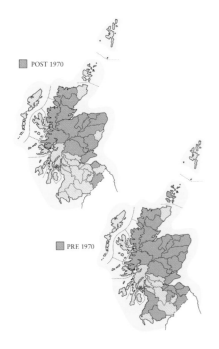

POST 1970

PRE 1970

11. *Goodyera repens* (L.) R. Brown

Synonym: *Satyrium repens* L.

Creeping Lady's-tresses

This late summer-flowering orchid is one of the few British species which is almost exclusive to Scotland. *Goodyera repens* is mainly restricted to the remnants of the ancient Caledonian Pine Forest in Strathspey and Cairngorm regions of the Highlands. It can also be found in pine plantations, especially mature ones that provide a semi-open ground cover habitat. *G. repens* is the only Scottish orchid with evergreen leaves and a creeping habit. The stalked ovate leaves, which usually have a distinct net-veined pattern, are arranged in a rosette which becomes less apparent as the flower stalk lengthens. Vegetative reproduction occurs by means of runners which spread through the pine needle litter.

Flowering plants range from 8–35cm tall. The upper stem and the outside of the bracts have dense glandular hairs; the ovaries are less hairy. Each inflorescence consists of a single spiral row of creamy white, sweetly scented flowers which are twisted so that they face more or less the same direction. The sepals have white glandular hairs on the outside and are blunt-tipped, as are the petals. The lateral sepals are spreading but the dorsal one forms a hood together with the petals. The lip is pointed at the tip and sac-like at the base. The margins of the anther are orange. The bracts are narrow and 10–15mm long, exceeding the length of the ovary. Pollination is thought to be by bumble-bees. *G. repens* can be found in flower during July and early August.

G. repens grows in moss and pine needle litter in forest clearings and fringes of mature pine woods. It is also occasionally found in more open areas among tall bell heather (*Erica cinerea*) but always with pines nearby.

In Scotland *G. repens* occurs mainly in the north and east-central Highlands. It is absent from the offshore islands, central and southwest Scotland. Although a record from the 1950's exists for Orkney, the population there is now thought to be extinct. In England, *G. repens* occurs in the northernmost counties, and in pine plantations in north Norfolk. It is not clear whether the latter are indigenous or were possibly introduced with pine seedlings from Scotland. It has not been recorded from Wales or Ireland. *G. repens* is widely distributed throughout north, central and western Europe and North America where it extends southwards to North Carolina.

No hybrids have been recorded for this species.

Roxburghshire, 15 vii 1990

Roxburghshire, 22 vii 1989   Roxburghshire, 15 vii 1990

Height: 2–10cm

Flowers:
Colour: yellowish green
Number: up to 20
Arrangement: lax inflorescence; bracts narrow and pointed, about equalling the ovary
Sepals: largest of the perianth parts, ovate-lanceolate, yellowish green, laterals pointing upwards, dorsal pointing downward
Petals: linear-lanceolate, green, spreading and backward-curving
Lip: striped light and dark green, triangular, and upward pointing between the lateral sepals
Ovary: straight with a twisted stalk

Leaves: 2–3 (–4), pale green, elliptic to ovate and often fringed with tiny bulbils

Stem: hairless and angled

Flowering: July to September

POST 1970

PRE 1970

12. *Hammarbya paludosa* (L.) O. Kuntze

Synonyms: *Ophrys paludosa* L.; *Malaxis paludosa* (L.) Swartz

Mogairlean Bogaich
Bog Orchid

The bog orchid, *Hammarbya paludosa*, is the smallest of our Scottish orchids, the whole plant blending perfectly with the yellow-green of sphagnum moss in the acid bogs which are its favoured habitat. Because it is so inconspicuous, this species is probably the most overlooked, and hence under-recorded, of all our wild orchids. Flowering is notoriously erratic: some years only an occasional flowering plant appears where many had been found in previous seasons.

Plants, which are completely yellowish green, range in height from 2 to 10cm. Concealed in the moss at the base of each stem are two pseudobulbs, the lower of which dies back each season to be replaced by a new one. One or two basal leaves, which are more or less reduced to sheaths, cover the upper pseudobulb, and a little further up the stem are two or three, rarely four, small narrow elliptic to ovate leaves, each fringed at their margins with tiny bulbils which detach to form new plants. *H. paludosa* also produces large quantities of dust-like seed following pollination by small insects. Inflorescences have up to twenty minute green flowers in which, alone among Scottish orchids, the lip points upwards – i.e. the flowers are hyper-resupinate. The lip, which is smaller than the sepals and petals, is striped light and dark green. Flowering occurs from July through to mid-September.

Sphagnum moss saturated with water is the only habitat suitable for this species. In consequence it is usually found in sphagnum bogs but also occurs along stream edges and loch margins where the moss is present.

In Britain, *H. paludosa* has a decidedly northern distribution, and is almost exclusively found in wet, upland boggy areas with acidic ground water. There are Scottish records from most of the west and central vice-counties, including the Outer and Inner Hebrides, from Orkney and, rarely, from Angus, south Aberdeenshire, Dumfriesshire and Kirkcudbrightshire. In England and Wales, the species is restricted by the lack of suitable habitat. However, it occurs in Cornwall, Devon, west Wales, northern England, Norfolk and the New Forest area of Hampshire. Despite an apparently favourable climate in Ireland, *H. paludosa* is confined to a very few vice-counties. Worldwide, the distribution of *H. paludosa* is circumboreal: it occurs in North America, Asia, Scandinavia and most northern European countries.

No hybrids of *H. paludosa* have been recorded in Europe.

West Ross, 10 vii 1990

West Ross, 10 vii 1990

West Ross, 10 vii 1990

Height: 6–28cm

Flowers:
Colour: yellowish green tinged reddish, lip white spotted red
Number: 4–13
Arrangement: lax inflorescence of slightly pendulous flowers; bracts triangular, pointed, to 3 **x** 1mm
Sepals: strap-shaped, green tinged with reddish brown; dorsal forming hood with petals; laterals spreading downwards and incurved
Petals: smaller, strap-shaped, greenish, forming hood with dorsal sepal
Lip: shorter and broader than petals and sepals, shallowly 3-lobed, white with crimson spotting at base
Ovary: straight with twisted stalk; in fruit downward-pointing

Leaves: no green leaves; 2–4 sheathing scale-leaves

Stem: slender, hairless and yellowish green or reddish

Flowering: May to August

13. *Corallorhiza trifida* Chatelin

Synonym: *Corallorhiza innata* R. Brown

Coralroot Orchid

The only European representative of the genus, *Corallorhiza trifida* has its British stronghold in Scotland. Because these saprophytic plants are small, pale-coloured and somewhat insignificant, it is suspected they may be under-recorded. A recent survey of suitable habitats – dune-slacks and willow / alder carr – in east and central Scotland resulted in many new records, including some localities with as many as 500 flowering plants.

The yellowish green *C. trifida* ranges in height from 6 to 28cm. Plants on dune slacks tend to be shorter and to have rather reddish stems. There are no green leaves but two to four long, sheathing scale-leaves clasp the lower half of the hairless stem. Inflorescences are lax with up to 13 more or less downward facing flowers. The pale green sepals and petals are strap-shaped, the dorsal sepal and the petals form a hood while the lateral sepals are spreading and somewhat incurved. The lip is shallowly three-lobed and white with crimson spots towards the base. Pollination is by small insects, although self-pollination is not uncommon. Flowering starts in late May and continues until August.

In Scotland, the main habitat of *C. trifida* is dense willow / alder carr on raised mires and lowland loch margins. More rarely it grows on well established dune slacks well covered by creeping willow (*Salix repens*). A third habitat which occasionally supports colonies of *C. trifida* is moss and pine-needle litter in mixed pine and birch woodland, where *Goodyera repens* and *Listera cordata* may also occur.

C. trifida is recorded from woodlands in many east and central vice-counties from the Borders northwards to East Sutherland and East Ross. It is also, though rarely, found on dune slacks in the east. Elsewhere in Britain it is mainly restricted to Northumbria and Cumbria with occasional records from more southerly counties. It is not recorded from Wales or Ireland. The species is widespread in continental Europe being absent only from the Mediterranean region. It is also found throughout North America and Greenland.

No hybrids have been recorded, but a rare peloric variant, with all three inner perianth parts resembling the lip has been recorded in Ayrshire and Moray.

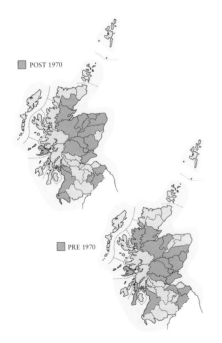

POST 1970

PRE 1970

Roxburghshire, 2 vi 1990

Selkirkshire, 1 vi 1991

The short red stems of *Corallorhiza*
trifida are indicative of dune slack
habitats. Fife, 8 vi 1989

Roxburghshire, 29 v 1990

Height: 4–20(–34)cm

Flowers:
Colour: greenish or green tinged with red or purple-brown
Number: 5–25
Arrangement: loose, cylindrical inflorescence; bracts long and pointed, length may vary markedly between plants
Sepals: short, broad and forming a loose hood with the petals
Petals: narrow, often almost hidden by the converging sepals
Lip: strap-shaped, 3-lobed, the two side-lobes longer than the mid-lobe; usually folded under flower
Spur: c.2mm, almost globular
Scent: slightly honey scented

Leaves: 3–6 ovate to oblong, blunt leaves towards base of stem, becoming narrower and more pointed further up

Stem: slightly angled, hairless and often tinged red or purplish brown

Flowering: June to early August

Mycorrhizal associate: unidentified species of *Ceratobasidium* isolated from population in Sutherland; *Thanatephorus orchidicola* isolated from another population in Sutherland

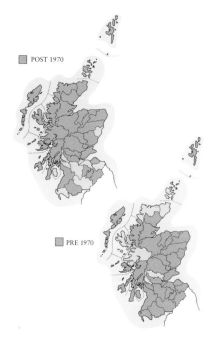

POST 1970

PRE 1970

14. *Coeloglossum viride* (L.) Hartman

Synonyms: *Satyrium viride* L.; *Habenaria viridis* (L.) R. Brown

Mogairlean Losgainn
Frog Orchid

Coeloglossum viride is usually between 10 and 20cm tall although its full size range is from 4cm to more than 30cm. The stem is angled, often reddish above, and has one or two brown basal sheaths. There are 3–6 leaves which are often concentrated towards the base of the stem. The lower sheathing leaves are spreading, 3–10cm long, ovate to oblong, and blunt at the tips. Further up the stem, they become narrower, more pointed, and non-sheathing. Between five and 25 flowers are arranged in a loose, cylindrical inflorescence. The bracts are green or tinged reddish and pointed; their length varies but the lower ones are much longer than the flowers. Flower colour ranges from green to green tinged with red or purple-brown. Sepals are short, ovate and broad, converging with the narrower petals to form a loose hood. The long strap-shaped lip is three-lobed with the side-lobes longer than the mid-lobe, and is usually folded back under the flower. The spur is about 2mm long and almost globular. Flowering is from June to early August.

Coeloglossum viride occurs in mountain grassland, on grassy ledges or in upland flushes, at up to 1,000m. It also grows in coastal grassland, dunes or the machair. The dune grassland habitats include many golf courses in Scotland among which those at North Berwick and Crail are home to many thousands of frog orchids.

The frog orchid is commoner in Scotland, where it is currently recorded from all but ten vice-counties, than in England and Wales. However, even in Scotland it can only be described as local or, at best, locally abundant. Nevertheless, numbers fluctuate from year to year and can increase at a surprising rate. For example, in 1956 only one plant was found on the whole island of Tiree, but by 1973 over 1000 plants were recorded from twelve different localities. In England and Wales *C. viride* is widespread but local, being absent from parts of Wales, the midlands and the extreme south west. It is also widespread in Ireland and, indeed, throughout most of north central Europe. The species also extends east to the Caucasus, Western Siberia, and as far as Turkestan and Kashmir.

In Scotland *C. viride* is reported to have hybridised with *Dactylorhiza purpurella*, *D. fuchsii* subsp. *hebridensis*, *Gymnadenia conopsea* and *Platanthera bifolia*. A long-bracted form of *C. viride* is sometimes recognised as a distinct variety, *C. viride* var. *bracteata*, but, as the length of the bracts is variable throughout the range of the species the variety is not universally accepted.

Fife, 16 vii 1990

Fife, 5 vii 1991

Fife, 5 vii 1991

Subsp. *conopsea*

Height: 15–40cm, rarely taller

Flowers:
Colour: rosy pink, or rarely white
Number: 20–50, rarely more
Arrangement: cylindrical inflorescence of small flowers; bracts green or green tinged purple, as long as flowers
Sepals: dorsal sepal held erect; lateral sepals 5–6mm long, with pointed tips, spreading and angled slightly, c.30°, downwards
Petals: forming loose hood
Lip: without distinct shoulders, 3-lobed; lobes almost equal
Spur: long, slender and curved
Ovary: $^1/_2$–$^2/_3$ the length of the spur
Scent: sweet with acidic overtones

Leaves: 3–5 sheathing leaves, narrow and lanceolate; 2–3 smaller non-sheathing leaves, narrow and pointed

Stem: hairless, purplish above

Flowering: June to August

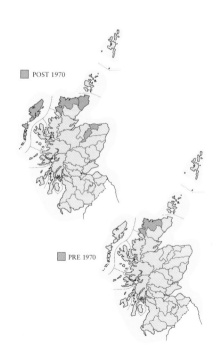

POST 1970

PRE 1970

15. *Gymnadenia conopsea* (L.) R. Brown

Lus Taghte
Fragrant Orchid

In Britain the fragrant orchid occurs as three subspecies: subsp. *conopsea*, subsp. *densiflora* and subsp. *borealis*. All three occur in Scotland but *G. conopsea* subsp. *borealis* is by far the most common. The spreading lateral sepals and long slender spur of *G. conopsea* make the species relatively easy to identify in the field; confusion with other Scottish orchid species is unlikely.

The three subspecies may be distinguished by the size and shape of the lip, and by the shape and position of the lateral sepals. Each has a characteristic scent which can be used as an aid to identification, but its intensity varies according to the weather conditions and the age of the flower - newly opened flowers have the strongest scent. Subsp. *conopsea* prefers dry calcareous grassland and dry limestone areas, subsp. *densiflora* is found in damper, base-rich places, while subsp. *borealis* is the least habitat-specific, being found in dryish hill-pasture, heathland and roadside verges. A fourth subspecies, subsp. *insulicola*, with dull reddish purple flowers and an unpleasant scent resembling rubber has been described from the Outer Hebrides. No current information is available relating to this subspecies, which may well have been *G. conopsea* subsp. *borealis*. Orchid species which are reported to have hybridized with *G. conopsea* (probably subsp. *borealis*) in Scotland, are *Coeloglossum viride*, *Dactylorhiza purpurella*, *D. fuchsii*, *D. maculata* subsp. *ericetorum* and *Pseudorchis albida*.

White-flowered forms of all three subspecies of fragrant orchid are known, and can be fairly common within some populations.

15a. *Gymnadenia conopsea* (L.) R. Brown subsp. *conopsea*

Synonyms: *Orchis conopsea* L.; *Habenaria conopsea* (L.) Bentham, non Reichenb. fil.

Plants of *Gymnadenia conopsea* subsp. *conopsea* normally range in height from 15–40cm but rare specimens up to 60cm occur. The hairless stem is often tinged purplish near the top and has two or three brown basal sheaths. There are three to five folded, narrow-lanceolate, bright green leaves near the base and, further up, two or three narrow non-sheathing leaves lying close to the stem. The cylindrical inflorescence consists of numerous small, pink, long-spurred flowers with a sweet, but faintly acidic scent. The lip is divided into three almost equal lobes; the pointed lateral sepals are 5–6mm long, spreading, and angled slightly downwards. Flowering is at its peak in June, the same as subsp. *borealis* but slightly earlier than subsp. *densiflora*.

This subspecies thrives on calcareous grassland and dry limestone areas. It is extremely rare in Scotland, the only records being from west Sutherland, Caithness, Banffshire and the Outer Hebrides. There is an unsubstantiated report from Orkney. It is much more common in southern England and Wales on dry chalk and limestone but is frequently recorded northwards to Durham. In Ireland, it is considered frequent in the central counties, and local elsewhere though hybrids have obscured its pattern in the west. The subspecies is frequently found on chalk and limestone hills in central Europe and extends to Siberia, China, Korea and Japan.

No hybrids of *G. conopsea* subsp. *conopsea* have been noted from Scotland, although it has been recorded as hybridising with a number of other orchids elsewhere.

Sutherland, 1 vii 1991

Sutherland, 1 vii 1991

Sutherland, 28 vii 1991

Height: 30–60cm, rarely taller

Flowers:
Colour: mauve-pink to purple pink
Number: up to 100
Arrangement: cylindrical inflorescence; bracts narrow and pointed, as long as the flowers, tinged purplish
Sepals: dorsal projecting forwards; laterals 6–7mm long, parallel-sided, blunt-tipped, and held horizontally
Petals: forming a loose hood
Lip: with distinct shoulders, 3-lobed, often with a pale whitish central patch, side-lobes larger than mid-lobe
Spur: slender and curved, 10–16mm long
Ovary: $^2/_3$ to $^3/_4$ the length of the spur
Scent: spicy sweet

Leaves: 3–5 sheathing leaves towards the base, lanceolate, rather broader than those of subsp. *conopsea* and subsp. *borealis*; non-sheathing leaves narrower and more pointed

Stem: robust, hairless and tinged purple near the top

Flowering: July to August

POST 1970

PRE 1970

15b.　*Gymnadenia conopsea* (L.) R. Brown subsp. *densiflora* (Wahlenberg) Camus, Bergon & A. Camus

Synonym: *Orchis conopsea* L. var. *densiflora* Wahlenberg

Gymnadenia conopsea subsp. *densiflora* is the most robust fragrant orchid, with plants up to 75cm tall being recorded, although typical plants are usually between 30 and 60cm tall. The scent is spicy sweet, with no acidic overtones. The two side-lobes of the lip are much larger than the mid-lobe and have distinct 'shoulders'. The lateral sepals have blunt tips, are 6–7mm long, and are held horizontally. In Scotland this subspecies flowers later than the other two, being at its best during July and early August.

Subspecies *densiflora* usually grows in calcareous fens and marshes although plants have been found thriving in other habitats such as wet ditches and, occasionally, north-facing chalk grassland. The only recent Scottish records are of plants in a damp field in West Ross and an unsubstantiated report of plants growing in a wet flush in upland grassland near Newcastleton in the Borders. However, a recent study of herbarium specimens in the Royal Botanic Garden Edinburgh has revealed several old specimens which had been identified only as *G. conopsea* from Forfarshire, Lanarkshire, Lothians, Peeblesshire and Berwickshire. In England, subsp. *densiflora* is largely restricted to fens and marshes in the south and west though it also occurs in chalk grassland on north-facing slopes in Sussex. There are records from Anglesey in north Wales and also from west Ireland. The subspecies is found throughout Europe and in some areas, such as the Dordogne, it is quite common in fens and other damp places. However, it is most common in the south east, notably in the Balkans.

No hybrids have been recorded in Scotland though they occur elsewhere in its range.

Both *Gymnadenia conopsea* subsp. *densiflora* and *Trollius europaeus* favour wet grassland sites. West Ross, 9 vii 1990

West Ross, 9 vii 1990

West Ross, 9 vii1990

Height: 15–25(–30)cm

Flowers:
Colour: dark pink to lilac
Number: 20–30+
Arrangement: cylindrical inflorescence; bracts pointed, tinged purplish and equalling the length of the flowers
Sepals: dorsal upright, laterals spreading, oval-lanceolate, pointed at tips, 4–5mm long and angled downwards
Petals: forming a loose hood
Lip: without shoulders, small, narrow, 3-lobed, side-lobes shorter than mid-lobe, lobes not always distinct
Spur: 11–14mm long, slender and curved
Ovary: $^1/_2$ to $^2/_3$ length of spur
Scent: very sweet, like carnations

Leaves: 3–5 near base long and narrow; the upper 2–3 narrower and more pointed

Stem: robust, hairless, upper part occasionally tinged purple

Flowering: June to August

15c. *Gymnadenia conopsea* (L.) R. Brown subsp. *borealis* (Druce) F. Rose

Synonyms: *Habenaria gymnadenia* Druce var. *borealis* Druce; *Gymnadenia conopsea* (L.) R. Brown var. *borealis* (Druce) Godfery; *G. conopsea* (L.) R. Brown var. *insulicola* Heslop-Harrison

Gymnadenia conopsea subsp. *borealis* is the most common fragrant orchid in Scotland, and is one of our most appealing native orchids. The number of flowering plants fluctuates widely from year to year, possibly due to changes in rainfall pattern.

Plants range between 15 and 30cm tall. The stem has three to five lanceolate leaves near the base which tend to be folded and arranged in two vertical ranks. In addition there are two or three narrow, pointed, non-sheathing leaves along the stem. The flowers are an unusual dark pink or lilac and carnation-scented. The lateral sepals are shorter and broader than those of the other subspecies. They are 4–5mm long, oval-lanceolate, pointed at the tip, spreading, and angled downwards. The dorsal sepal is upright and the petals form a loose hood. Unlike the other subspecies the lip of subsp. *borealis* is small and the side-lobes, which may sometimes be indistinct, are shorter than the mid-lobe. The spur is slender, curves downwards, and is almost twice the length of the ovary. Flowering is from June until August.

G. conopsea subsp. *borealis* has been recorded from every vice-county in Scotland. It grows in a wide variety of habitats including hill pasture, flushed *Molinia* grassland, heathland and roadside verges. It is also found in the north of England, Sussex, Hampshire and Cornwall, as well as in Wales and the west of Ireland. Elsewhere in Europe it occurs in upland areas but its distribution needs further study.

Outside Scotland, hybrids have been recorded with subsp. *conopsea* and subsp. *densiflora*.

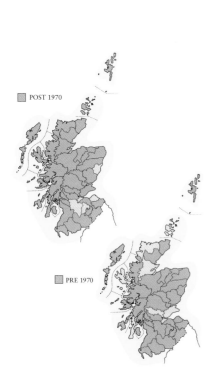

POST 1970

PRE 1970

Skye, 13 vi 1990

Selkirkshire, 16 vii 1989

Roxburghshire, 26 vi 1990

Height: 10–30cm

Flowers:
Colour: creamy or greenish white
Number: 10–60, usually 30–40
Arrangement: dense cylindrical inflorescence of tiny flowers; bracts small, pointed, as long as ovary
Sepals: small, elliptical, forming a loose hood with petals
Petals: small, elliptical
Lip: deeply 3-lobed, mid-lobe longer than side-lobes
Spur: thick and conical, c. 2–3mm long; curving downwards
Scent: faint vanilla fragrance

Leaves: usually four oblong to oblanceolate sheathing leaves; 1–2 more pointed, non-sheathing leaves

Stem: hairless

Flowering: June to July

16. *Pseudorchis albida* (L.) A. & D. Löve

Synonyms: *Satyrium albidum* L.; *Habenaria albida* (L.) R. Brown; *Gymnadenia albida* (L.) L. C. M. Richard; *Leucorchis albida* (L.) E. Meyer

Mogairlean Bàn Beag
Small-white Orchid

The small-white orchid is one of only two species of *Pseudorchis* found in Europe. Though often occurring singly or in small numbers, this attractive species is quite widespread and it is worth looking out for, especially in the West Highlands and the Inner Hebrides.

P. albida ranges in height from 10 to 30cm. The stem is hairless with two or three pale sheaths at the base. There are normally about four oblong to oblanceolate, sheathing leaves on the lower part of the stem, and above one or two narrower, lanceolate, non-sheathing leaves. Bracts are lanceolate and as long as the ovary. The dense inflorescence has numerous tiny creamy or greenish white, bell-shaped flowers which droop slightly and face to one side. In each flower, small sepals and petals form a broad, loose hood over a deeply 3-lobed lip which has a small, thick, conical, down-curving spur at its base. The flowers have a delicate vanilla fragrance and are at their best during June and July.

The small-white orchid usually grows on well drained upland pasture and grassy mountain ledges but near the west coast it occasionally occurs at sea-level. It has been found growing amongst recently burned *Calluna vulgaris* on grouse moorland but the orchid becomes less vigorous and eventually disappears as the ling regrows.

In Scotland *P. albida* is recorded from almost every vice-county although there are no confirmed records from the Outer Hebrides, and very few from the southwest. Despite its widespread distribution, it is not numerous occurring in only one or two localities in many vice-counties. It is found in northern England, west Wales and Ireland, and is widespread throughout Europe from the mountains of Spain eastwards to the CIS and northwards to Scandinavia and Iceland, where it is extremely common.

A hybrid with *Gymnadenia conopsea* has been recorded from a number of Scottish locations, and one with *Dactylorhiza maculata* subsp. *ericetorum* has been recorded from Orkney.

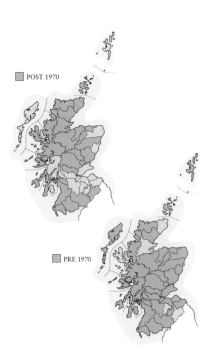

POST 1970

PRE 1970

Roxburghshire, 20 vi 1990

Roxburghshire, 29 vi 1989

Height: 20–40cm, rarely to 60cm

Flowers:
Colour: greenish white
Number: 10–30, rarely more
Arrangement: lax inflorescence; bracts narrow, pointed, about as long as ovary
Sepals: lateral sepals ovate, spreading and pointed, dorsal sepal smaller and blunt
Petals: narrow, forming a semi-circular hood over the pollinia
Lip: 10–16mm long, strap-shaped, tapering towards the blunt tip; distinctly greenish, particularly towards the tip
Spur: to 25mm long or more, slender and curved, often strongly so; entrance clearly seen
Scent: heavy fragrance, most evident at night
Pollinia: prominent and divergent

Leaves: normally two large lower leaves, sub-opposite, elliptical or elliptic-lanceolate and blunt; 1–5 smaller leaves along stem, narrower and more pointed

Stem: angled and hairless

Flowering: May to July

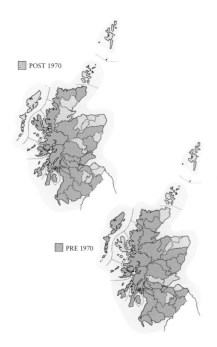

POST 1970

PRE 1970

17. *Platanthera chlorantha* (Custer) Reichenbach

Synonyms: *Orchis chlorantha* Custer; *Habenaria chlorantha* (Custer) Babington

Mogairlean an Dealain-dè Mòr
Greater Butterfly-orchid

The best way to differentiate between the greater and lesser butterfly-orchids is to examine the structure of the flower. In *P. chlorantha* the pollinia diverge so that the entrance to the spur can be clearly seen. In *P. bifolia* by contrast the pollinia are close together and parallel, making the entrance to the spur less obvious.

Plants of the greater butterfly-orchid normally range between 20 and 40cm tall but occasionally reach 60cm. The hairless angled stem has one to three brown basal sheaths. The lower leaves are opposite or sub-opposite and elliptical to elliptic-lanceolate with blunt tips; the one to five upper leaves are smaller and more pointed. The lax inflorescence is more or less pyramidal with 10–40 large, greenish white, strongly scented flowers in which the lateral sepals are ovate, pointed and spreading while the dorsal sepal is smaller and blunt-tipped. The petals are somewhat smaller than the sepals and form a semi-circular hood over the pollinia. At the base of the narrow, strap-shaped lip is a long, slender, curving spur up to 25mm or more. The pollinia are divergent and the entrance to the spur can be clearly seen. *P. chlorantha* is pollinated by night-flying moths and flowers from May to July.

The most common habitat for the greater butterfly-orchid is grassland pasture but it can be found in many habitats including damp, deciduous woodland or scrub, marshy areas, and even heath moorland.

Although Scottish populations of *P. chlorantha* are rarely large, distribution is widespread, with records from most vice-counties. It is most frequently recorded from the west; currently it is not known from Orkney, Shetland, the Outer Hebrides and some eastern vice-counties. In the rest of the British Isles, it is much more common, occurring in every vice-county in England and Wales and in many of the limestone areas in the west of Ireland. Elsewhere in Europe, it is equally widespread, occurring mainly on calcareous soils in open woodland, scrub or pasture.

Although *P. chlorantha* and *P. bifolia* may occur together and flower at the same time, hybrids between the two are rare, presumably because differences in the position of the pollinia make cross-pollination unlikely. Nevertheless hybrids have been recorded from Scotland and elsewhere in Britain. Rare variants with unusual flower structure are also known. One from Mid Perthshire was originally thought to be an intergeneric hybrid between *P. chlorantha* and *Pseudorchis albida* but this has since been discounted.

Roxburghshire, 29 vi 1989

Roxburghshire, 29 vi 1989

Roxburghshire, 29 vi 1989

Height: 15–30cm, rarely to 45cm

Flowers:
Colour: white, slightly tinged yellowish green
Number: normally 15–25, rarely more
Arrangement: woodland form has loose inflorescence, moorland form more compact and cylindrical; bracts narrow and pointed, slightly shorter than ovary
Sepals: laterals spreading and lanceolate; dorsal erect and more triangular
Petals: forming a broad, loose, triangular hood
Lip: 6–12mm long, strap-shaped
Spur: 13–23mm long; slender
Scent: heavy, sweet fragrance
Pollinia: vertical, parallel and close together

Leaves: usually two lower leaves, sub-opposite, elliptical; 1–5 upper leaves, narrower and more pointed

Stem: angled and hairless

Flowering: woodland form May to June; moorland form June to July

Mycorrhizal associate:
Tulasnella calospora isolated from population in Aberdeenshire

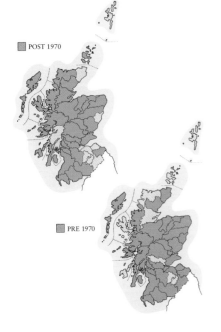

POST 1970

PRE 1970

18. *Platanthera bifolia* (L.) L. C. M. Richard

Synonyms: *Orchis bifolia* L.; *Habenaria bifolia* (L.) R. Brown

Mogairlean an Dealain-dè Beag
Lesser Butterfly-orchid

Two distinct morphological forms of the lesser butterfly-orchid occur in Scotland, each being found in different habitats. Plants growing on moorland are short with a dense inflorescence and broad pointed leaves, whereas those in woodland are usually taller with the flowers spaced further apart and longer, narrower, less pointed leaves. Both forms have a narrow column and parallel pollinia.

Plants of the lesser butterfly-orchid range from 15 to 45cm. At the base of the stem are one, two or three brown or whitish sheaths, with two large sub-opposite, elliptical leaves just above. Further up the stem there are between one and five smaller, more pointed leaves. The inflorescence is either dense or lax, depending on which form is found, with up to 25 large (11–18mm across), white flowers, tinged yellowish green. The lateral sepals are lanceolate and spreading while the dorsal sepal is upright and triangular. Like the greater butterfly-orchid, each flower has a long, slender spur, 13–23mm long but the pollinia are parallel and close together. The woodland form flowers between late May and June; the moorland form continues from June well into July. Pollination is by night-flying moths.

The woodland form can tolerate shade and is usually found in damp woods or scrub where it often grows through a ground cover of moss. This form is most common in the beech woods of southern England and records thin out northwards. It is reliably recorded from only a few Scottish localities, mainly in deciduous woods in the Borders. However, since many records of *P. bifolia* do not differentiate between the two forms, the woodland form may be under-recorded. In Scotland and northern England, the moorland form is the predominant one. It prefers damp heather moorland with slightly acidic soil but also tolerates drier tussocks in marshy ground. All but ten of the Scottish vice-counties have records of this form and it also occurs on the moors of northern England and Wales, in damp pastures and heaths in Ireland, and further south on the moors of Devon and Cornwall.

Taking both forms together, *P. bifolia* is probably more common in Scotland than *P. chlorantha*. In Europe in general the lesser butterfly-orchid is probably as widespread as the greater butterfly-orchid and has a similar distribution.

In 1949 a hybrid with *Coeloglossum viride* was recorded on South Uist but has not been re-discovered. Hybrids between the two British *Platanthera* species are rare, probably because of differences in flower morphology and peak flowering time.

North Uist, 15 vi 1990

North Uist, 15 vi 1990

North Uist, 15 vi 1990

Woodland form from birch/willow carr.
Berwickshire, 2 vii 1989

Height: 6–15cm

Flowers:
Colour: normally intense violet-purple but can be pink
Number: 3–12
Arrangement: loose inflorescence; bracts somewhat membranous, flushed purple, shorter than ovary
Sepals: oval; dorsal forming a hood with petals; laterals heavily green-veined
Petals: narrower than sepals
Lip: broader than long and shortly 3-lobed to almost entire, mid-lobe shorter or no longer than reflexed side-lobes, with paler dark-spotted central area
Spur: as long as ovary, held horizontally or upward, slightly enlarged at the tip
Scent: sweetly scented but very variable in intensity

Leaves: rosette of up to 7 or 8 bluish green elliptical to lanceolate basal leaves; 1–2 smaller leaves further up the stem, pointed and clasping

Stem: hairless, flushed purple

Flowering: May to June

19. *Orchis morio* L.

Green-winged Orchid

The green-winged orchid is well named from the distinctive green veins on the wing-like lateral sepals. One of the more restricted of Scotland's orchids, it can be rather easily confused with the commoner early purple orchid. Although *O. morio* seems to prefer a more open situation in short grazed turf than the sheltered ledges and gullies favoured by the early purple orchid, the two can occur in the same locality. Unfortunately *O. morio* has suffered loss of habitat in recent years following road improvements in Ayrshire, the only area where it is currently known.

In Scotland, *O. morio* is a small plant ranging from about 6 to 15cm high. It has several narrow basal leaves and one or two pointed clasping leaves further up the stem. The flowers are usually purple but pink forms also occur. They have a loose hood formed from the dorsal sepal and narrow petals while the two spreading lateral sepals have the distinctive green veins. The lip is three-lobed and has a pale central area marked with dark purple dots. A broad, horizontal or upward-pointing spur is present. The flowers of some of the Ayrshire plants are scented but many have no discernible smell. Pollination is by insects, especially bees. In Scotland, flowering is early, taking place during May and June.

The only documented Scottish localities for this species are in Ayrshire on rough pasture on old sandstone cliff tops or overlaying basaltic lavas. There is a reliable record from 1958 of a single flowering plant found near Tomintoul in the Grampians, but older records from other localities are rather dubious. Some of them were probably misidentified plants of *Orchis mascula*.

Elsewhere in the British Isles the green-winged orchid is more common, particularly in southern England, Wales and the central counties of Ireland. It is recorded from most central and southern countries of continental Europe, but is replaced in the Mediterranean region by a number of subspecies.

Although no hybrids have so far been recorded in Scotland *O. morio* has reportedly hybridised elsewhere in Britain with *O. mascula*. Pink-flowered forms occur, but white-flowered plants have not been recorded from Scotland.

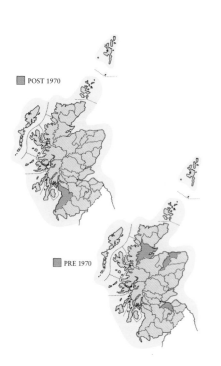

POST 1970

PRE 1970

Ayrshire, 17 v 1991

Ayrshire, 17 v 1991

Ayrshire, 19 v 1989

Height: c.8–46cm, rarely more

Flowers:
Colour: purple-pink, sometimes pale
Number: 20–50, rarely more
Arrangement: loose inflorescence; bracts somewhat membranous, as long as ovary and flushed purple
Sepals: oval and pointed, laterals angled upwards and outwards; dorsal forming loose hood with the petals
Petals: more blunt than sepals
Lip: broad, 3-lobed, side-lobes more or less reflexed, mid-lobe longer, with paler spotted central area and often with distinct notch
Spur: stout and slightly upturned
Scent: often smells of tom-cats, occasionally pleasantly scented

Leaves: basal rosette of 4 to 8 oblong, blunt tipped leaves, usually heavily marked with elongated dark purple blotches; two or three smaller more pointed clasping leaves further up the stem

Stem: hairless, flushed purple above

Flowering: late April to early July

POST 1970

PRE 1970

20. *Orchis mascula* (L.) L.

Synonym: *Orchis morio* L. var. *mascula* L.

Moth-ùrach
Early Purple Orchid

The early purple orchid, *Orchis mascula*, is one of the first orchid species to flower, blooming as early as April. It has been well documented for many years under a variety of common (sometimes coarse) names. John Lightfoot in his *Flora Scotica* of 1777 has a particularly fine description of the orchid which he identifies as 'long-purples' or 'dead men's fingers' used for Ophelia's garland in Shakespeare's Hamlet. Lightfoot also uses the name 'male fool-stones' for *O. mascula* and gives a detailed account of its use as an aphrodisiac.

O. mascula ranges from 8 to 46cm tall and has a stout stem with a basal rosette of four to eight oblong leaves which usually have numerous dark purple blotches. Higher up the stem are two or three smaller clasping leaves which may also have purple spots. The loose inflorescence has between 20 and 50 purple-pink flowers, sometimes more on taller plants. Lateral sepals are angled upwards and outwards and, unlike *O. morio*, are not conspicuously veined. On the broad, purplish pink three-lobed lip is a paler central patch spotted with dark purple. The side-lobes are usually reflexed and are shorter than the mid-lobe which usually has a distinct notch. The stout, blunt spur is often upturned. Flowering occurs from late April at sheltered coastal sites, to as late as July on grassy ledges in the highlands.

O. mascula prefers base-rich soils, and is primarily found in open, deciduous woodland, damp rich grassland and grassy coastal braes. However, it can also be found in other open habitats, such as upland pasture, roadside verges, and even on high mountain ledges where the soil is not too acid. As one of our commonest orchids it can be found in every vice-county in Scotland. In England it grows on chalk and limestone grassland and in deciduous woodland. In Ireland it is a notable feature of the spring vegetation of the Burren. It is locally abundant throughout the rest of Europe.

No hybrids have been reported for Scotland, but elsewhere hybrids with *O. morio* have been found. It is not uncommon to find a pale-flowered form of *O. mascula*, but white-flowered forms have not been reported in Scotland.

East Lothian, 20 v 1991

Berwickshire, 13 v 1989

Subsp. *fuchsii*

Height: 17–40 (–70)cm

Flowers:
Colour: pale to deep pinkish purple, rarely white, marked with darker lines or dashes
Number: 30–70
Arrangement: elongated, tapering to cylindrical inflorescence; bracts pointed, longer than ovary
Sepals: laterals spreading; dorsal forming loose hood with petals
Petals: forming loose hood with dorsal sepal
Lip: deeply 3-lobed, 8–12mm wide, the mid-lobe triangular and longer than the rhomboid side-lobes; marked with symmetrical pattern of lines and dashes
Spur: straight, slender, 1–2mm wide
Scent: faint fragrance

Leaves: 5–10, lowest is broadest and rounded at tip, next two are longest, remainder becoming narrower, shorter and more pointed; usually marked with transversely elongated blotches

Stem: hairless

Flowering: late May to late July

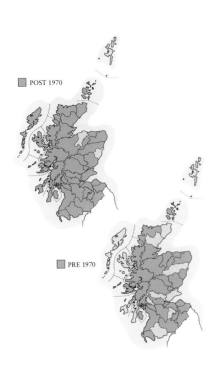

POST 1970

PRE 1970

21. *Dactylorhiza fuchsii* (Druce) Soó

Common Spotted-orchid

Dactylorhiza fuchsii, which is locally abundant in Scotland, has three subspecies: subsp. *fuchsii*, subsp. *hebridensis*, and subsp. *okellyi*.

D. *fuchsii* ranges between 6 and 70cm tall with between five and ten sometimes slightly folded leaves. The lowest leaf is rounded at the tip, wider, and shorter than the next two or three. The remainder become shorter, narrower and non-sheathing with distance up the stem. Usually the leaves have transversely elongated blotches, some being heavily marked. The pointed bracts are almost always longer than the ovary and less than 3mm wide at the base. Numerous white to deep lilac-pink flowers make up the inflorescence which varies in shape from compact to elongate and from cylindrical to tapering. With the exception of subsp. *okellyi* sepals and petals are marked to some extent with darker lines and dots. The dorsal sepal forms a loose hood with the petals and the lateral sepals are spreading. A deeply 3-lobed lip has the mid-lobe longer than the side-lobes and is marked with a symmetrical pattern of darker lines and dashes in subsp. *fuchsii* and *hebridensis*. A straight spur, 1–2mm in diameter, is usually half to two-thirds the length of the ovary. This species flowers from late May to late July and is pollinated by bees and small insects lured by nectar produced inside the spur. When identifying white-flowered plants of D. *fuchsii* differences in lip shape and leaf width are diagnostic. Occasional white-flowered plants in a colony are more likely to be a colour variant of that subspecies whereas a group of white-flowered plants in isolation could be subsp. *okellyi*.

As with most of the dactylorchids, hybridisation is quite common and interspecific hybrids have been recorded between the common spotted-orchid and a number of the closely related species which are found in Scotland. These include D. *maculata* subsp. *ericetorum* and D. *purpurella*, together with a number of subspecies of D. *incarnata*. Intergeneric hybrids have also been described with *Coeloglossum viride*, and *Gymnadenia conopsea*.

21a. *Dactylorhiza fuchsii* (Druce) Soó subsp. *fuchsii*

Synonyms: *Orchis fuchsii* Druce; *Dactylorchis fuchsii* (Druce) Vermeulen

This subspecies is the commonest of the three. Plants are usually robust, varying in height from 17 to 70cm with an average around 40cm. Lower leaves are elliptic to broad lanceolate; the upper ones become lanceolate. Usually all are marked with transversely elongated blotches. The dense, many-flowered inflorescence becomes cylindrical as the flowers open. Darker lines and dots occur on the sepals while the lip has a double loop of broken lines and dashes. The mid-lobe of the lip is triangular and longer than the side-lobes.

The common spotted-orchid can be found in a variety of habitats though it prefers calcareous or neutral soils. Although absent from most of our heather moorland areas, it can be found in occasional alkaline flushes where it may grow alongside its more acid-loving relative, the heath spotted-orchid. D. *fuchsii* subsp. *fuchsii* is generally found in open woodland, coastal or upland pasture and road verges. In recent years the undisturbed habitats provided by verges along motorways and major trunk roads have enabled this subspecies to increase considerably in numbers. It is recorded from almost every vice-county in Scotland including the Hebrides though the dominant form there is subsp. *hebridensis*. In many areas the common spotted-orchid can be described as locally abundant.

D. fuchsii subsp. *fuchsii* is common in chalk and limestone areas of southern Britain. It thrives in Ireland and parts of Wales, and is widespread and common in western and central Europe.

While variants with pure white flowers are not common in this subspecies, forms with almost white flowers can be found. However, close examination will generally reveal a hint of very pale pink lip marking. It is with these almost white-flowered plants that confusion with *D. fuchsii* subsp. *okellyi* may occur, but lip shape and leaf width are distinguishing features.

Argyll, 28 vi 1990

Roxburghshire, 2 vii 1989

Midlothian, 6 vii 1990

Midlothian, 6 vii 1990

Height: 6–30cm, averaging about 20cm

Flowers:
Colour: rose-pink to deep lilac, often pale magenta, rarely white
Number: up to 40
Arrangement: blunt-topped compact inflorescence; bracts pointed, longer than ovary
Sepals: laterals spreading; dorsal forming loose hood with petals
Petals: forming loose hood with dorsal sepal
Lip: deeply 3-lobed, 10–15mm wide, the triangular mid-lobe longer than the rhomboid-shaped side-lobes; marked with symmetrical pattern of lines and dashes
Spur: straight
Scent: faint fragrance

Leaves: 5–10, lowest is normally broadest, next two are longest, remainder becoming narrower, shorter and more pointed; purple-blotched or spotted

Stem: hairless

Flowering: June and July

POST 1970

PRE 1970

21b. *Dactylorhiza fuchsii* (Druce) Soó subsp. *hebridensis* (Wilmott) Soó

Synonyms: *Orchis hebridensis* Wilmott; *Dactylorchis fuchsii* subsp. *hebridensis* (Wilmott) Heslop-Harrison; *Dactylorhiza hebridensis* (Wilmott) Averyanov

Urach-bhallach
Hebridean Spotted-orchid

D. fuchsii subsp. *hebridensis* is the most striking of all the spotted orchids, especially when it grows in large numbers in its typical habitat, the machair of the Outer Hebrides. This small subspecies more or less replaces subsp. *fuchsii* in the Outer Isles. It usually grows to only about 20cm and has fewer, normally spotted or blotched, leaves than subsp. *fuchsii*. The flowers are larger, normally a much deeper magenta pink, and arranged in a compact inflorescence.

This subspecies is virtually confined to machair grassland and short coastal turf, in which it can occur in very high numbers. Subspecies *hebridensis* only occurs in the British Isles and grows chiefly in the Outer Hebrides but has also been recorded from Jura, Tiree, Shetland and a few mainland locations on the north west coast. Elsewhere it is known from a few coastal areas in the west of Ireland and also from Cornwall.

D. fuchsii subsp. *hebridensis* is known to have hybridised with *Coeloglossum viride*, *D. maculata* subsp. *ericetorum*, *D. purpurella*, and a subspecies of *D. incarnata*, probably subsp. *coccinea*.

Lewis, 14 vii 1991

Lewis, 14 vii 1991

Lewis, 14 vii 1991

Height: 12–20cm, rarely taller

Flowers:
Colour: white, occasionally faintly marked
Number: 20–30
Arrangement: flat-topped inflorescence; bracts narrow and pointed, longer than ovary
Sepals: laterals spreading; dorsal forming hood with petals
Petals: forming hood with sepals
Lip: 3-lobed, all ± equal in length, mid-lobe triangular, side-lobes rounded with strongly curving outer margins; white, unmarked or faintly marked
Spur: straight, c. 1–2mm diameter
Scent: usually fragrant

Leaves: fewer and narrower than in subsp. *fuchsii*; all slightly glaucous, narrow, unmarked or faintly spotted and with broadly rounded tips

Stem: hairless

Flowering: June to July

POST 1970

PRE 1970

21c. *Dactylorhiza fuchsii* (Druce) Soó subsp. *okellyi* (Druce) Soó

Synonyms: *Orchis okellyi* Druce; *Dactylorchis fuchsii* subsp. *okellyi* (Druce) Vermeulen; *Dactylorhiza maculata* (L.) Soó subsp. *okellyi* (Druce) H. Baumann & Kunkele; *D. okellyi* (Druce) Averyanov

Plants of *D. fuchsii* subsp. *okellyi* found in Scotland are usually between 12 and 20cm, slender, with narrow unspotted leaves and white, often fragrant, flowers. The flowers are smaller than those of the other two subspecies and the three lobes of the lip are almost equal in length with the side-lobes rounded rather than rhomboid in shape. It is said that single white-flowered plants are unlikely to be subsp. *okellyi* since it is usually found in colonies.

The main habitat for *D. fuchsii* subsp. *okellyi* is calcareous grassland and it is chiefly found in north and west Ireland. Only a few Scottish records are known from Kintyre and S Ebudes with an old unsubstantiated record from Sutherland. There is a single record from the Isle of Man. It does not occur elsewhere in Europe.

No hybrids or varieties have been recorded in Scotland.

Kintyre, 21 i 1993

Kintyre, 17 vi 1991

Kintyre, 17 vi 1991

Height: 15–30cm, rarely to 50cm

Flowers:
Colour: usually pale pinkish lilac, rarely darker or white, marked with darker dots and lines
Number: 5–20, rarely more
Arrangement: ± pyramidal inflorescence; bracts longer than ovary
Sepals: long and narrow, laterals spreading; dorsal forming loose hood with petals.
Petals: forming loose hood with dorsal sepal
Lip: broader than long, 3-lobed, mid-lobe triangular and usually shorter than the two rounded side-lobes; marked with irregular dots and lines
Spur: 3–7 x 1mm, shorter than ovary

Leaves: 2–4 narrow lanceolate sheathing leaves in lower part of stem, lowest is usually shorter but not wider; 1–3 (rarely 4) smaller non-sheathing leaves further up the stem. All leaves usually marked with round pale purplish spots

Stem: hairless

Flowering: May to July, rarely August

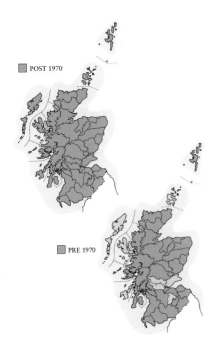

POST 1970

PRE 1970

22. *Dactylorhiza maculata* (L.) Soó subsp. *ericetorum* (E. F. Linton) P. F. Hunt & Summerhayes

Synonyms: *Orchis maculata* L. subsp. *ericetorum* E. F. Linton; *Dactylorchis maculata* (L.) Vermeulen subsp. *ericetorum* (E. F. Linton) Vermeulen; *Dactylorhiza ericetorum* (E. F. Linton) Averyanov; *D. maculata* (L.) Soó subsp. *rhoumensis* (Heslop-Harrison) Soó; *Orchis fuchsii* Druce subsp. *rhoumensis* Heslop-Harrison

Mogairlean Mòintich
Heath Spotted-orchid

The heath spotted-orchid is one of the two most common orchids found in Scotland, the other being *Dactylorhiza purpurella*, the northern marsh-orchid. Lip shape and marking are the best means of distinguishing the heath spotted-orchid from the otherwise similar common spotted-orchid, *Dactylorhiza fuchsii*. In the former the side-lobes are conspicuous and longer than the mid-lobe, whereas in *D. fuchsii* the mid-lobe is equal to or longer than the side-lobes. Irregular reddish dots and lines marking the lip of *D. maculata* subsp. *ericetorum* contrast with the more clearly defined symmetrical, looped lines on the lip of *D. fuchsii*. The heath spotted-orchid is a slender plant rarely exceeding 30cm tall – the rare exceptions require careful identification as some of the known hybrids are rather taller plants. Flower colour ranges from white to pinkish lilac, with pale colours being most common. A few flowers are arranged in a more or less pyramidal inflorescence. The lateral sepals are spreading with the dorsal sepal and petals forming a loose hood. Sepals and petals have darker markings. The lip is broader than long and shallowly three-lobed with the side-lobes broadly rounded and conspicuous, and are usually longer than the small triangular mid-lobe. The lip is marked with darker pink irregular dots and short lines. The spur is 3–7mm long, about 1mm wide, and untapered. Pollination is by bees and flies. Flowering extends from May until July, rarely to August.

Although the heath spotted-orchid normally prefers well drained acid moorland, it can occasionally be found in damper areas such as upland flushes, marshes or sphagnum bogs. Primarily a mountain species, it has been recorded growing at altitudes up to 915m in Scotland. However, owing to the rugged nature of much of our western coastline, it can also be found growing close to sea-level there. Open birch, alder, oak or pine woodland occasionally supports this species which is found in every vice-county in Scotland. It has a more scattered distribution in England and Wales, occurring on moorland and heaths as far south as Devon and Cornwall. It grows throughout the north and west of Ireland. Elsewhere in Europe the heath spotted-orchid is restricted to Sweden and the Netherlands.

Where the ranges of *D. purpurella* and *D. fuchsii* overlap with that of *D. maculata* subsp. *ericetorum*, there is a great deal of hybridisation. Indeed, the most common orchid hybrids found in Scotland are derived from these three species and in many areas where they overlap, hybrid plants are far more common than the parents. Since many of the hybrids show hybrid vigour, some large colonies contain a high proportion of tall, robust flower spikes, each displaying a range of characteristics inherited from the parent plants. An intergeneric hybrid with *Gymnadenia conopsea* occurs in the west of Scotland, and there is a record from Orkney of a hybrid with *Pseudorchis albida*. Hybrids with most of the subspecies of *D. incarnata* and *D. traunsteineri* are recorded from other parts of the British Isles. A subspecies, *D. maculata* subsp. *rhoumensis*, was described from Rhum in the Inner Hebrides, but is no longer considered to be distinct.

Skye, 13 vi 1990

Lewis, 14 vii 1991

Subsp. *incarnata*

Height: 7–20(–50)cm

Flowers:
Colour: flesh-pink with darker pink markings
Number: 10–40, usually 20–30
Arrangement: tight, dense inflorescence; bracts pointed, lower ones longer than the flowers
Sepals: laterals angled upwards, marked with dots and rings; dorsal forming hood with the petals
Petals: forming hood with dorsal sepal
Lip: shallowly 3-lobed with side-lobes strongly reflexed, marked with symmetrical loops enclosing an area of lines and dots
Spur: 6–9mm long, over 2mm diameter, stout and conical, downward pointing

Leaves: 3–5 sheathing leaves, erect, keeled and hooded at tips, 0–2 non-sheathing leaves; all yellowish-green, unmarked

Stem: stout, hairless, flushed purplish above

Flowering: May to late July

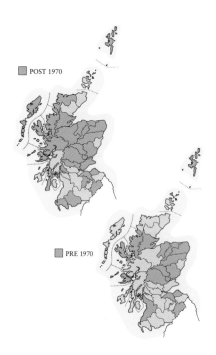

POST 1970

PRE 1970

23. *Dactylorhiza incarnata* (L.) Soó

Mogairlean Lèana
Early Marsh-orchid

In the British Isles *Dactylorhiza incarnata* is an extremely variable species with five subspecies, four of which occur in Scotland. Colours range from the bright crimson red of subsp. *coccinea*, to the flesh-pink of subsp. *incarnata* and the mauve-purple and magenta of subspp. *cruenta* and *pulchella*. In addition, many populations have plants with white flowers.

D. *incarnata* varies in height from seven to 30cm, rarely more. It has three to six sheathing leaves which are more or less erect, keeled, and hooded at the tips. Further up the stem there is usually one, rarely two, non-sheathing leaves but occasional plants have none. In all subspecies except subsp. *cruenta* the leaves are unmarked. The inflorescence is dense, slightly less so in subsp. *cruenta*, and the bracts, especially the lower ones, are long. The lip is not deeply lobed, and the sides are usually strongly reflexed, giving the lip a narrow appearance from the front. On the lip, darker markings make up a distinctive double loop enclosing an area of dots and lines. The spur is thick and conical; pollination is thought to be by bees. Subspecies of D. *incarnata* are primarily differentiated by flower colour, and to a lesser extent on habitat. Hybrids between the subspecies of D. *incarnata* and D. *purpurella*, D. *fuchsii*, D. *fuchsii* subsp. *okellyi* and D. *maculata* subsp. *ericetorum* have all been reported from Scotland.

23a. *Dactylorhiza incarnata* (L.) Soó subsp. *incarnata*

Synonyms: *Orchis incarnata* L.; *O. latifolia* (L.) Pugsley; *O. strictifolia* Opiz; *Dactylorchis incarnata* (L.) Vermeulen

The flesh-pink flowers of the early marsh-orchid are undoubtedly its most eyecatching feature. In Scotland plants usually range between seven and 20cm in height, rarely more. They have three to five yellowish-green sheathing leaves and usually one non-sheathing leaf. The former are lanceolate, keeled, narrowly hooded at the tip and are held more or less erect. All leaves are unmarked. Bracts are usually flushed reddish purple and the lower ones are much longer than the flowers. A tight cylindrical inflorescence is made up of small, pale flesh-pink flowers which have their lateral sepals erect and folded back above the loose hood formed by the petals and dorsal sepal. The flowers are otherwise as described for the species. D. *incarnata* subsp. *incarnata* flowers from mid-May to July.

In Scotland this subspecies is restricted to damp meadows, marshes and mountain flushes, preferring soil with an alkaline or neutral pH. In the recent past many such wetlands have been drained for agriculture or forestry and so the subspecies has probably declined. However, since most records are simply of 'D. *incarnata*', it is difficult to be sure of the distribution of this, or the other subspecies. It is probably fairly widespread but local throughout Scotland, occurring in most vice-counties except for a few in the north and west. It is the most frequent subspecies in south and southeast England, and occurs widely in continental Europe as far east as the Crimea.

Rhum, 26 vi 1989

Skye, 13 vi 1990

Height: 8–15cm, rarely to 30cm

Flowers:
Colour: deep crimson to madder red
Number: 10–40, usually about 30
Arrangement: dense cylindrical inflorescence; bracts pointed, lower ones longer than the flowers
Sepals: laterals angled upwards, marked with dots and rings; dorsal forming hood with petals
Petals: forming hood with dorsal sepal
Lip: shallowly 3-lobed with side-lobes strongly reflexed; darker red marking of distinct, symmetrical loops enclosing an area of lines and spots
Spur: 6–9mm long, over 2mm diameter, stout, conical, downward pointing

Leaves: 3–6 broad, keeled and hooded sheathing leaves towards the base of the stem; further up are 1–2 narrower, non-sheathing leaves; all dark green, unmarked

Stem: very stout, hairless, flushed purplish above

Flowering: June and July

23b. *Dactylorhiza incarnata* (L.) Soó subsp. *coccinea* (Pugsley) Soó

Synonyms: *Orchis latifolia* L. var. *coccinea* Pugsley; *Dactylorchis incarnata* (L.) Vermeulen subsp. *coccinea* (Pugsley) Heslop-Harrison; *Dactylorhiza coccinea* (Pugsley) Averyanov; *Orchis strictifolia* Opiz subsp. *coccinea* (Pugsley) Clapham

Dactylorhiza incarnata subsp. *coccinea* is perhaps the most striking of the four subspecies found in Scotland, especially when large numbers of them carpet damp, open dune slacks. Their colour has been variously described as vermilion, brick-red, crimson, ruby-red, deep indian-red, madder red, or even maroon. Whatever name is chosen, few other orchids can match it.

Subsp. *coccinea* is a shorter and much stockier plant than subsp. *incarnata*, rarely exceeding 30cm in height and normally ranging between 8 and 15cm. There are three to six broad, thick, dark green, keeled and hooded sheathing leaves towards the base and usually one or two narrower non-sheathing leaves further up the stem. All leaves are unmarked. The dense cylindrical inflorescence often gives plants a rather squat appearance; many smaller plants may appear stemless, as the inflorescence is almost concealed by the large enveloping leaves. Flowering is at its best during June and July.

Subsp. *coccinea* is found mainly on dune slacks but is also frequent on Hebridean machair. In addition to the Hebrides, it is found in Kintyre, Shetland and the east coast from Aberdeenshire southwards. It is almost exclusively coastal in Scotland though there are a few records of it occurring a short way inland, on Skye for example. This subspecies is also found in duneland habitats in England and Wales and the Isle of Man. In Ireland it occurs around the coast of Donegal and near inland loughs, and there are unconfirmed records from the Netherlands.

A hybrid plant between this subspecies and the rare *D. purpurella* subsp. *majaliformis* was recorded in 1985 on the Outer Hebrides.

POST 1970

PRE 1970

Lewis, 14 vii 1991

East Lothian, 16 vi 1989

North Uist, 14 vi 1990

North Uist, 15 vi 1990

Height: 10–30cm

Flowers:
Colour: mauve-purple or magenta with dark purple or red markings
Number: 10–30, rarely more
Arrangement: tight, dense inflorescence; bracts pointed, lower ones longer than the flowers
Sepals: laterals spreading back and upwards, marked with dots and rings; dorsal forming hood with the petals
Petals: forming hood with dorsal sepal
Lip: shallowly 3-lobed with side-lobes strongly reflexed; marking of distinct, symmetrical loops enclosing an area of lines and dots
Spur: 6–9mm long, stout and conical, downward pointing

Leaves: 4–5 sheathing leaves, erect and spreading, 1–2 non-sheathing leaves clasping the stem; pale green, narrow, not strongly keeled, all unmarked

Stem: stout, hairless, flushed purplish above

Flowering: May to late July

POST 1970

PRE 1970

23c. *Dactylorhiza incarnata* (L.) Soó subsp. *pulchella* (Druce) Soó

Synonyms: *Orchis incarnata* var. *pulchella* Druce; *Dactylorchis incarnata* (L.) Vermeulen subsp. *pulchella* (Druce) Heslop-Harrison; *Dactylorhiza pulchella* (Druce) Averyanov

D. incarnata subsp. *pulchella*, like subsp. *coccinea*, is known only from Britain and Ireland. Plants of subsp. *pulchella* are very similar to those of subsp. *incarnata* in all but flower colour, which is a mauve-purple or magenta with dark purple or red loop and line markings.

This subspecies prefers boggy areas with rather more acidic conditions than subsp. *incarnata*. At a number of localities both subspecies can grow side by side, particularly where alkaline flush areas occur in heather moorland. Subsp. *pulchella* is found in hills and mountains throughout the north and west of Scotland including some of the more rugged islands. It is also known from a few lowland areas in Scotland, particularly in the southwest mainland. Elsewhere in the British Isles, it is known from the New Forest in Hampshire, the hills of North Wales, and the counties of Donegal, Mayo and Sligo in Ireland.

A hybrid between subsp. *pulchella* and subsp. *cruenta* has been recorded.

Skye, 13 vi 1990

Stirlingshire, 13 vi 1992

Height: 7.5–20cm

Flowers:
Colour: pale lilac-purple to violet-magenta
Number: 5–20
Arrangement: somewhat lax inflorescence; bracts about as long as the flowers; heavily spotted and flecked or entirely stained dark purple
Sepals: laterals held erect, broad, blunt at tips marked with spots and rings; dorsal forming hood with petals
Petals: forming loose hood with dorsal sepal
Lip: not deeply lobed but mid-lobe usually prominent; side-lobes not strongly reflexed. Marked with symmetrical loops, dashes and dots
Ovary: marked with purplish flecks
Spur: thick at base, conical and curved

Leaves: usually 3–5 erect, lanceolate, keeled, sheathing leaves which are slightly recurved and hooded at the tip; sometimes 1 non-sheathing leaf present. All bright to yellowish green, spotted on both surfaces.

Stem: slender, hairless, flecked and stained pinkish violet in upper part

Flowering: mid- to late June sometimes extending into July

POST 1970

23d. *Dactylorhiza incarnata* (L.) Soó subsp. *cruenta* (O. F. Mueller) P. D. Sell

Synonyms: *Orchis cruenta* Mueller; *Dactylorchis incarnata* subsp. *cruenta* (Mueller) Vermeulen

Flecked Early Marsh-orchid

The main character which sets *D. incarnata* subsp. *cruenta* apart from other marsh-orchids is the presence of spots on both sides of the leaves. There are small and large violet brown blotches on the upper surface, rather more dense towards the tip, and several smaller paler spots on the lower surface. Other distinguishing features are the shape of the lip although the patterning is of typical *D. incarnata* symmetrical loops, dashes and dots, and the purplish streaks and flecks on the upper stems, bracts and ovaries.

In Scotland, this subspecies is only known in two small areas of West Ross. These have a neutral to slightly alkaline pH and are similar to the species-rich flush communities elsewhere in the north and west Highlands, including those where two other rare Scottish orchids, *D. traunsteineri* and *D. lapponica*, have been found. The Irish distribution is more widespread with most records from County Clare and other areas in the south and west, although its numbers are decreasing. Subsp. *cruenta* is not known in England or Wales but occurs in the Swiss, Austrian, German and French Alps and most of Scandinavia.

A hybrid between *D. incarnata* subsp. *cruenta* and subsp. *pulchella* has been recorded.

West Ross, 16 vi 1991

West Ross, 16 vi 1991

West Ross, 16 vi 1991

Height: 5–10(–15)cm

Flowers:
Colour: violet-purple
Number: 5–15
Arrangement: dense, short and cylindrical inflorescence; lower bracts slightly longer than flowers; all flushed purple
Sepals: elliptical, laterals spreading upwards and outwards; dorsal curved forward and forming a loose hood with the petals
Petals: forming loose hood with dorsal sepal
Lip: broader than long, distinctly 3-lobed, mid-lobe longer than the spreading side-lobes; line and loop markings darker purple
Spur: about as long as ovary, cylindrical, conical, downward pointing

Leaves: 3, sometimes 4, broad pointed sheathing leaves crowded towards base of stem; 1 smaller non-sheathing leaf; all leaves bright green, the upper surfaces heavily blotched, sometimes almost entirely coloured brownish purple, often purple-edged

Stem: short, sturdy, grooved and hairless

Flowering: late May to mid-June

POST 1970

24. *Dactylorhiza majalis* (Reichenbach) P. F. Hunt & Summerhayes subsp. *scotica* E. Nelson

Dactylorhiza majalis subsp. *scotica* is the only subspecies of *D. majalis* to occur in Scotland, and it grows nowhere else. The only confirmed records are from a small area of the North Uist coast. A number of plants from north and west Scotland, originally thought to be *D. majalis* subsp. *occidentalis*, have been transferred to other taxa. Subsp. *scotica* is quite distinctive: its short squat stature and conspicuously blotched leaves set it apart from any of the other closely related dactylorchids.

It normally ranges from 5 to 10cm tall but occasional plants may reach 15cm. There are three, occasionally four, broad basal leaves with the lowermost one rather shorter than the others and also a small non-sheathing leaf higher up the stem. The leaves are often purple-edged and the basal ones are heavily blotched, sometimes to the extent of being almost completely purple-brown. Bracts and upper stem are suffused purple. There is a dense inflorescence of violet-purple flowers with thick, downward pointing spurs. Lateral sepals are angled upwards, spreading and forward pointing and the 3-lobed lip is marked with dark purple lines and loops. A loose hood is formed by the dorsal sepal and petals.

Subspecies *scotica* grows on areas of machair intermediate between the wet hollows favoured by *D. incarnata* subsp. *coccinea* and the small, drier hummocks colonised by *D. purpurella* subsp. *purpurella*. It is known from two strands on North Uist although apparently suitable stretches of machair occur throughout the western isles. There is also an unconfirmed report from Jura. The reasons for this remarkably restricted distribution are unknown.

Plants have been found which appear to be intermediate between *D. majalis* subsp. *scotica* and *D. purpurella* subsp. *purpurella*.

North Uist, 14 vi 1990

North Uist, 14 vi 1990

North Uist, 14 vi 1990

Subsp. *purpurella*

Height: 10–30cm, rarely taller

Flowers:
Colour: bright-purple to magenta with dark purple dash or dot markings
Number: 10–40+
Arrangement: dense, flat-topped inflorescence, 3–7cm or more long; bracts narrow, pointed, unspotted and stained purple
Sepals: laterals erect with darker markings; dorsal forming loose hood with petals
Petals: forming loose hood with dorsal sepal
Lip: ± flat, broadly diamond-shaped or obscurely three-lobed, bright red-purple with dark purple broken line markings
Spur: 6–9mm x 2mm, cylindrical and conical; slightly downward pointing

Leaves: 4+ sheathing leaves, 5–12cm long, broad, usually unmarked but occasionally with fine spots towards tips; 1–2 non-sheathing leaves

Stem: ribbed, usually robust, hairless

Flowering: late May to July

Mycorrhizal associate: *Thanatephorus cucumeris* and an unidentified *Rhizoctonia* species isolated from populations in Aberdeenshire

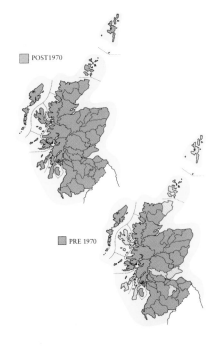

POST 1970

PRE 1970

25. *Dactylorhiza purpurella* (T. & T. A. Stephenson) Soó

Mogairlean Purpaidh
Northern Marsh-orchid

The northern marsh-orchid is one of the two most common orchids found in Scotland, the other being *D. maculata* subsp. *ericetorum*, the heath spotted-orchid. *D. purpurella* has a north British distribution with another species, *Dactylorhiza praetermissa*, replacing it in the south. In Scotland it occurs as two subspecies, *purpurella* and *majaliformis*. The former is widespread in damp places, while the latter is rare and very habitat-specific, being found on the north and northwest coast of the mainland and in the Hebrides.

    D. purpurella is usually 10–35cm tall and robust. The four or more broad, glaucous green sheathing leaves are generally unmarked in subsp. *purpurella* and marked with round purple blotches in subsp. *majaliformis*. Flower colour provides another distinction, being bright purple to magenta in subsp. *purpurella* and more pinkish purple in subsp. *majaliformis*. Sepals of both forms are angled upwards. The lip is not deeply lobed and may even be diamond-shaped, with heavy, but not continuous, marking.

    D. purpurella not only hybridizes freely with a number of other dactylorchids but also with other genera. There are known hybrids with *D. maculata* subsp. *ericetorum*, *D. fuchsii*, subspecies of *D. incarnata*, *Coeloglossum viride*, and subspecies of *Gymnadenia conopsea*.

25a. *Dactylorhiza purpurella* (T. & T. A. Stephenson) Soó subsp. *purpurella*

Synonyms: *Orchis purpurella* T. & T. A. Stephenson; *Dactylorchis purpurella* (T. & T. A. Stephenson) Vermeulen; *Dactylorhiza majalis* subsp. *purpurella* (T. & T. A. Stephenson) D. M. Moore & Soó

D. purpurella subsp. *purpurella* is an opportunistic orchid that can be found in large numbers growing in a variety of habitats, including urban sites. One reason for its success may be the large quantities of seed that are regularly produced.

    It is a rather robust plant, normally 10–30cm high, but taller in some conditions – in damp open woodlands for example. The four or more broad, lanceolate sheathing leaves are usually unmarked although rarely they may have small purple spots near the tips. There are one or two smaller non-sheathing leaves further up the stem. Lower bracts are about as long as the flowers. The dense, flat-topped inflorescence has 10–40, sometimes more, distinctive bright-purple to magenta flowers. The lateral sepals are angled upwards, often touching back to back. There is a weakly lobed, diamond-shaped lip marked with short dark purple lines and dots. The thick, conical spur is shorter than the ovary. Flowering is from late May to late July.

    The northern marsh-orchid can be found in a wide variety of habitats but is most common in marshes and damp flushes where the ground water has a neutral pH. It can also be found in large numbers in damp ditches, dune slacks, road verges and undisturbed gardens. There are even records from relatively dry habitats such as rubble on derelict urban sites and troughs in alpine gardens.

    In Scotland *D. purpurella* subsp. *purpurella* is found in every vice-county though there are no recent records from Renfrewshire. Elsewhere in Britain it flourishes in north and west England, north Wales and Northern Ireland, but outwith the United Kingdom it is only known in Norway, Sweden and the Faroe Islands.

Roxburghshire, 17 vi 1989

Roxburghshire, 17 vi 1989

Height: 15–35cm, rarely taller

Flowers:
Colour: pinkish purple with darker purple lip markings
Number: 15–30+
Arrangement: dense, cylindrical inflorescence, usually shorter than that of subsp. *purpurella*; lowest bracts as long as flowers, usually spotted or faintly suffused purple
Sepals: laterals angled upwards and spreading, purple-marked; dorsal forming loose hood with petals
Petals: elliptical and pointed, narrower than sepals and forming a loose hood with dorsal sepal
Lip: more 3-lobed than that of subsp. *purpurella*; marked with darker purple
Spur: broad, conical and slightly downward pointing

Leaves: 4 or more broad sheathing leaves heavily marked on the upper surface with large dark violet spots or blotches; 1–2 non-sheathing leaves, also spotted; a few plants have unmarked leaves

Stem: robust and ribbed, hairless, slightly suffused purple near the top

Flowering: May to July

POST 1970

25b. *Dactylorhiza purpurella* (T. & T. A. Stephenson) Soó subsp. *majaliformis* E. Nelson

Dactylorhiza purpurella subsp. *majaliformis* was previously misidentified as *D. majalis* subsp. *occidentalis*, but the only form of *D. majalis* now recognised in Scotland is subsp. *scotica*.

Subspecies *majaliformis* can be most easily distinguished from subsp. *purpurella* by its leaves and bracts which are heavily spotted or blotched. However, there are also differences in flower size, shape and colour. Those of subsp. *majaliformis* have larger more pinkish purple petals and sepals than subsp. *purpurella* and the lip of the former is more distinctly 3-lobed. Plants of subsp. *majaliformis* also tend to be rather taller, up to 35cm, and more robust than specimens of subsp. *purpurella*. Confusingly however, a small proportion of plants in Caithness populations of subsp. *majaliformis* may be small and weak with pale violet flowers. Plants with unmarked leaves also occur in the same locality; these have rather distinctive glaucous green leaves.

The only known habitats for this subspecies are close to the sea, often on seaward-facing grassy slopes. It is recorded from the Outer Hebrides, Kintyre and the west and north coasts of mainland Scotland. Elsewhere it is only known in North Denmark.

Caithness, 1 vi 1990

Caithness, 1 vi 1990

Height: 6–18cm, rarely to 24cm

Flowers:
Colour: magenta-purple or magenta-red, rarely lilac
Number: 3–12, rarely more
Arrangement: lax inflorescence; bracts lanceolate, purple-flushed, margins often tinged purple, invariably some spots on either or both surfaces
Sepals: laterals erect and blunt, marked with dark rings, elongated spots and dots; dorsal forming loose hood with petals
Petals: forming loose hood with dorsal sepal
Lip: 3-lobed, mid-lobe longer than side-lobes and often reflexed at tip; strongly marked with lines, rings and dots
Spur: robust, ± cylindrical and usually straight

Leaves: 2–3 spreading sheathing leaves; 0–2 non-sheathing leaves, all spotted and blotched, sometimes purple-edged; very rarely unmarked.

Stem: rather slender, hairless, upper part often suffused purplish

Flowering: late May to July

Mycorrhizal associate: material examined in 1992 showed the presence of typical orchid mycorrhiza but identification was not possible

POST1970

26. *Dactylorhiza lapponica* (Hartman) Soó

Synonyms: *Orchis angustifolia* M. Bieb. var. *lapponica* Hartman; *Dactylorhiza pseudocordigera* (Neumann) Soó; *D. traunsteineri* (Sauter ex Reichenbach) Soó subsp. *lapponica* (Hartman) Soó

Lapland Marsh-orchid

Dactylorhiza lapponica has only recently been recognised in Scotland. Plants of this species were previously thought to be *D. traunsteineri*, and then transferred to *D. majalis* subsp. *occidentalis*. Only in 1988 was it realised that they belonged to *D. lapponica*, a species new to the British Isles.

The Lapland marsh-orchid only grows in base-rich hill flushes. Plants are 6–18cm tall, rarely more, and rather slender. They have 2–3 oblong-lanceolate sheathing leaves, which are heavily marked on the upper surface with dark purple-brown spots and blotches which are often transversely elongated; they may also be purple-edged. Plants with unmarked basal leaves occur but are extremely rare. There are up to two erect, non-sheathing leaves which are flushed and edged purple, spotted on the upper surface, and sometimes also with a few small flecks on the lower surface. The bracts too are flushed, edged and spotted with purple. They are lanceolate with the lower ones as long as the flowers and the upper ones shorter. About 12 flowers usually make up the lax inflorescence. Flower colour is magenta-purple to magenta-red, rarely lilac. The lip is strongly marked with intense dark violet-purple or dark crimson lines, rings and dots, occasionally merging in the central part to form a dark patch. The mid-lobe is longer than the side-lobes and the pointed tip is often reflexed. The spur is robust and more or less cylindrical. Time of flowering varies appreciably with climatic conditions from late May to July.

In Scotland *D. lapponica* can be found in base-rich hill-flush communities from just above sea-level on South Harris and Rhum to 300m at some localities on the west coast of the mainland. It is unknown elsewhere in the British Isles but can be found in Scandinavia and the French, Swiss and Italian Alps.

Although there are no confirmed records of hybrids with *D. lapponica*, at one locality in the west where it grows together with *D. traunsteineri* and flowers at the same time, one or two plants have been found with intermediate characters.

Inverness-shire, 30 v 1990

Inverness-shire, 3 vi 1991

An unusually-marked form.
Rhum, 22 vi 1989

Inverness-shire, 3 vi 1991

Height: 9–20cm, rarely more

Flowers:
Colour: magenta-purple to magenta-pink or lilac with deeper markings
Number: up to 20, rarely more
Arrangement: lax inflorescence; lower bracts longer than flowers, narrow and pointed, flushed purplish, unspotted
Sepals: laterals spreading, pointed, sometimes faintly marked; dorsal forming loose hood with petals
Petals: forming loose hood with dorsal sepal
Lip: 3-lobed, with slightly reflexed side-lobes; mid-lobe usually pointed with reflexed tip, longer than the side-lobes; marked with dots and lines
Spur: stout, blunt and conical

Leaves: 2–3, rarely more, sheathing leaves well-spaced along the stem, second lowest usually the longest; linear-lanceolate, bright light green and almost always unmarked; 0–1 erect non-sheathing leaves; unmarked and flushed purplish

Stem: slender, hairless upper part often suffused purplish

Flowering: late May to mid-June

POST 1970

PRE 1970

27. *Dactylorhiza traunsteineri* (Sauter ex Reichenbach) Soó

Synonyms: *Orchis traunsteineri* Sauter ex Reichenbach

Narrow-leaved Marsh-orchid

As with other members of this difficult group of dactylorchids, there has been a history of misidentification of *Dactylorhiza traunsteineri* in Scotland. It was first confirmed near Loch Maree in 1983. Several additional populations have since been found on base-rich hill flushes in the west of Scotland; the habitat favoured by two other rare orchid species, *D. lapponica* and *D. incarnata* subsp. *cruenta*.

Plants of *D. traunsteineri* are slender and usually 15–20cm tall, with two or three (rarely up to five) long, narrow, pointed sheathing leaves. A smaller, erect non-sheathing leaf may also be present. The leaves are a bright light green and almost always unmarked. The bracts, which are also unmarked, are narrow and pointed, the lower ones longer than the flowers. Bracts, the non-sheathing leaf and the upper stem are flushed pinkish purple. The inflorescence is lax and fairly narrow, bearing up to 20 pale magenta or pinkish purple flowers. These have erect, spreading lateral sepals which are unmarked or faintly marked. The dorsal sepal forms a hood with the pointed petals. The mid-lobe of the lip is much longer than the side-lobes, narrow, and usually pointed with a reflexed tip. Unlike *D. lapponica*, the dots and lines marking the lip are not intensely coloured, nor are they violet.

In Scotland *D. traunsteineri* invariably occurs in damp flushes over or near natural springs with calcareous ground water. This habitat type is similar to that favoured by *D. lapponica*, and at one locality in west Scotland the two have been found growing side by side.

It is a rare species in Scotland, only known from a few colonies in the Inner Hebrides and on the west coast of the mainland. They are in West Ross, Applecross, Ardnamurchan, Kintyre, and Tiree. However, it seems likely that other localities remain to be discovered. Elsewhere in Britain, *D. traunsteineri* occurs in two bands, one from Yorkshire to North Wales and the other from East Anglia across to the Bristol area. There are a dozen or so localities scattered through Ireland. The orchid is also found in Scandinavia and throughout mainland Europe from France to Poland and Czechoslovakia.

Hybrids are known to occur with *D. fuchsii*, *D. maculata* subsp. *ericetorum* and *D. incarnata* but no confirmed records exist for Scotland. However in Ardnamurchan where both this species and *D. lapponica* grow together, a possible hybrid with characteristics from both was found recently.

Inverness-shire, 3 vi 1991

Inverness-shire, 3 vi 1991

Inverness-shire, 3 vi 1991

Height: 11–50cm, rarely more

Flowers:
Colour: bright pink, rarely white
Number: (30–) 40–80
Arrangement: dense pyramidal inflorescence, becoming more cylindrical when mature; bracts narrow and pointed, lower ones as long as or longer than ovary
Sepals: laterals broadly lanceolate, curved and spreading; dorsal curving forward to form a hood with the petals
Petals: forming hood with dorsal sepal
Lip: wedge-shaped, deeply 3-lobed, with two plate-like ridges at base
Spur: c. 12mm long, slender, longer than the ovary
Scent: slightly unpleasant foxy smell

Leaves: 3–4 narrow lanceolate sheathing leaves towards the base of the stem; up to 5 or 6 smaller erect pointed leaves along the stem; all grey-green

Stem: slender, hairless

Flowering: June to August

28. *Anacamptis pyramidalis* (L.) L. C. M. Richard

Synonym: *Orchis pyramidalis* L.

Mogairlean nan Coilleag
Pyramidal Orchid

Anacamptis pyramidalis is confined to a few isolated localities in Scotland. It ranges from 11 to 50cm high with three or four narrow lanceolate basal leaves that curve away from the stem and six or so upper leaves that become progressively smaller and more erect further up the stem. The basal leaves develop in autumn and winter. All leaves are grey-green. The flowers are usually bright pink, rarely pure white, have a foxy smell and form a tight, compact inflorescence. There are three, more or less equal, well-defined lobes to the lip and also two raised, plate-like ridges at the base which guide insects towards the spur entrance. The spur is very long and straight. *A. pyramidalis* flowers from late June until early August.

In Scotland this distinctive orchid is to be found among marram (*Ammophila arenaria*) and lyme grass (*Elymus arenarius*) on dunes, in calcareous grassland or occasionally along disused railway embankments. It has been known to have persisted in one east coast locality in high numbers for over 100 years. On the Scottish mainland it occurs at a few isolated localities in the Borders, Fife, Kintyre and on the south west coast. There are also a few records from the Hebrides. In England and Wales this orchid is found in most vice-counties, and it is also widespread in Ireland and throughout the rest of Europe. It also occurs in some Middle Eastern countries and in North Africa.

No hybrids have been found in Scotland but an intergeneric hybrid with the fragrant orchid, *Gymnadenia conopsea*, is recorded from England.

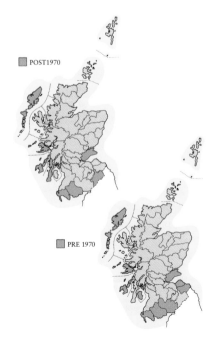

POST 1970

PRE 1970

Wigtownshire, 6 vii 1991

Wigtownshire, 6 vii 1991

One of a small colony growing on a
disused railway embankment.
Roxburghshire, 2 vii 1989

7. ORCHID
Photography

Orchid Photography

Sidney Clarke

*Writing field notes
Isle of Barra, Outer Hebrides
4 viii 1992*

The photographs illustrating this book were taken with either a Mamiya 645 super (6 x 4.5cm) 120 roll film camera or a Contax 35mm camera. The techniques that follow are based on these but most modern single lens reflex cameras (SLRs) will give good results. The most important requirements are accurate viewing and focusing and the facility of interchangeable lenses and close-up devices. Most modern SLRs also incorporate some form of exposure metering. For plant photography only two metering 'modes' are generally used, namely 'manual', where the photographer has full control over shutter speed and aperture settings, and 'aperture priority', where the aperture is chosen and the camera automatically sets the correct shutter speed.

A camera chosen specifically for plant photography should ideally incorporate:

(a) Interchangeable focusing screens

The standard screen supplied by most manufacturers normally incorporates a microprism ring or split-image wedges in the centre of the field. These are not suitable for critical focus when working at close range especially when using extension tubes or bellows. The best all-round screen for plant photography has as its centre field an area of fine ground glass. Although not as fast to use as the microprism or split-images for distant subjects, for close-up work and when using long telephoto lenses it is infinitely more useful.

(b) The facility of attaching a right-angle finder to the eyepiece of the viewfinder pentaprism

This allows the photographer to use the camera close to the ground to view the subject without having to lie on sometimes wet grass. As the right angle finder can be used with the camera supported in either the horizontal or the vertical position, it is more convenient than the interchangeable waist level finders that are available for some cameras. (For most plant portraits the camera will be used vertically.)

(c) Mirror lock

This is useful for close-up and macro photography where one can, after focusing and setting the controls, lock the mirror in the 'up' position to minimise vibration induced by the moving mirror. This is especially relevant to roll film cameras.

(d) 'Spot' metering

This is something of a misnomer as true spot meters measure an angle of around 1°. However, there are several cameras on the market where a small central area can be used for metering. This is usually referred to as 'spot metering' to differentiate it from the averaging or centrally weighted systems found in most cameras. Being able to choose the precise area from which to take the exposure reading is an extremely valuable aid to plant photography.

The standard lens supplied with most cameras has an angle of view of around 45°. The focal length for 35mm format is 50mm and that for the roll film format is 80mm. This standard lens is suitable for larger plants but, as it rarely focuses closer than 45cm, it is not suitable on its own for small plants. The least expensive method of focusing closer is to add close-up lenses.

These come in various dioptres (strengths) and are simply screwed into the filter thread of the standard lens. A set of three of these lenses would be required to cover a reasonable close-up focusing range.

An alternative, but more expensive, way of achieving closer focus with a standard lens is to use it in conjunction with one or more extension tubes. These are placed between the camera body and lens. They usually come in three different lengths and may be used singly or in combination to achieve the image size required.

The main drawback to using close-up lenses or tubes is that there can be 'dead' areas where sharp focus cannot be achieved. It can also be time consuming trying to find which of the set will give the desired image size. Moreover, the standard camera lens is optically corrected to give its best results when focused at infinity. At closer distances optical aberrations can affect the result.

The ideal solution is to replace the standard lens with a macro lens which will allow focusing down to half life size (with some macro lenses even to life size). Macro lenses are optically corrected for close-up work. If restricted to only one macro lens, the most useful is a 50–60mm lens with a minimum aperture of f/22. Wide angle lenses are particularly useful for habitat shots and those that will focus close can produce extremely effective 'plant in habitat' pictures. Moving closer to the plant produces a larger image of it while the lens, which will have an angle of around 60°–75° depending on the exact focal length, includes a good sweep of the background. Wide angle lenses of 28 or 35mm focal length (50mm for roll film cameras) are ideal. These are also very useful for working in confined places, such as cliff ledges, where movement is restricted.

Long focus lenses are necessary for situations where the subject cannot be approached closely. These lenses are available as macro lenses, usually with focal length of 90 or 100mm, from several manufacturers. Long macro lenses come into their own when working from a tripod on steep slopes. As they have a narrower angle of view (around 24°) than the standard lens or normal focal length macro, they can be used to good effect to exclude unwanted distracting objects from the background.

The tripod is probably the most under-estimated piece of equipment in the plant photographer's outfit. Most brands of cameras and lenses will nowadays give good results, but there are few tripods on the market that will completely satisfy the plant photographer's requirements.

The first requirement is stability. Secondly it should allow use of the camera as close to the ground as possible. It is this second requirement that rules out many otherwise excellent tripods. The tripod needs a good camera mounting head to allow as much flexibility in framing the subject as possible. A three-way head enabling separate adjustment in each direction is worth the extra expense.

The only other pieces of equipment required for daylight photography are a cable release (to release the camera shutter without jarring) and a lens hood (to prevent stray light falling on the front element of the lens and spoiling the image with 'flare').

Although filters are seldom required for colour photography of plants there are exceptions. At high altitude and near the sea, results can be improved by reducing ultraviolet light with a 'skylight' filter. Overcast conditions often produce excess blue light and a 'warming' filter (No. 81) can give a more pleasing colour balance to the final photograph. A polarising filter will help reduce surface reflections from shiny leaves or from water surfaces. It is also the only filter which can be used with colour film to darken a blue sky when photographing habitats.

The best all round colour slide films for photographing orchids by daylight are those with a speed of around 50 ASA (now termed ISO = International Standards Organisation), although films of 100 ASA can be used especially in formats larger than 35mm. Slower films, of around 25 ASA, will give superior results but only on windless, bright days. These conditions are not often met in Scotland so the slightly faster film is the more practicable. High speed films of around 200 ASA or faster may seem to be an ideal solution for windy days, but this is not so since speed is traded off against quality. The photographs illustrating this book have been shot on either Kodachrome 64 or Fujichrome Velvia.

Many otherwise perfect plant portraits are spoiled by their backgrounds and so it is worthwhile, if you have a choice, considering this in selecting a specimen to photograph.

The background should be as unobtrusive as possible. Light coloured stones and dead grasses should carefully be removed as they are sure to produce fussy backgrounds. Dead grasses lying near to the horizontal can be very distracting when out of focus. Include a pair of scissors in your kit for this tidying up, but be careful not to overdo it.

The most suitable form of illumination is soft daylight, preferably on a windless day! Strong sunlight should be avoided if at all possible as it produces a contrast range that will be beyond the capabilities of the film to fully record. Soft lighting will give detail in all parts of the image and will give more choice in selecting a suitable background, as the absence of shadows will allow the camera to be placed on any side of the plant. If photographs have to be taken when the sun is shining, the sunlight should be filtered through some kind of diffusing material to create a similar effect to soft daylight. Diffusers can be improvised from various materials such as tissue paper or white nylon and there is a range of commercially available products.

Flash lighting should be regarded as a last resort. It can be effective when employed with experience and creative flair, but poor flash photographs of plants are all too common. Some of the photographs in this book have been taken with the use of flash – not by choice but as a matter of necessity. The photography for this book was done to a strict timetable over three growing seasons and weather conditions were not always kind when the plants were at their best!

It is important in a portrait to get more or less the whole plant in sharp focus. As all Scottish orchids are upright in habit, this can only be achieved by supporting the camera in a position where its back is parallel to the flower stem. With the smaller species such as *Hammarbya paludosa*, the camera needs to be at ground level, hence the importance of choosing a suitable tripod.

Having positioned the camera, focus on the feature you regard as most critical. Usually this will be the flower head. Next choose the smallest aperture that you consider usable under the prevailing lighting conditions: use at least f/16, preferably f/22, or f/32 if you can. The smaller the aperture that you choose, the greater will be the depth-of-field and hence more of the plant will be within the range of acceptable sharpness. A small aperture will require the use of a fairly slow shutter speed, usually somewhere between $^1/_{15}$ sec. and $^1/_4$ sec.

If it is windy do not be tempted to use a high shutter speed and wide aperture. The depth-of-field will not be sufficient even if the high speed has frozen the moving plant. It is better to set the camera controls for the best results, wait patiently for a lull in the wind, and release the shutter using a cable release the instant the plant is motionless.

Achieving the correct exposure for colour slide film is important and can be tricky if you do not have a camera with spot metering. A 'washed out' subject is the all too common result of the camera or hand-held exposure meter reading light reflected from the background and surroundings as well as the flower. With automatic cameras one can compensate for this by setting the compensation control to give $^1/_2$ to 1 stop less than the indicated exposure. The exact amount will depend on the variation in brightness across the field of view.

An alternative method for automatic or manual exposure cameras is to take a substitute reading from a 'grey card' which can be obtained from most photographic shops. Grey cards, usually 8" x 10", have a grey side with a reflectance value of 18% representing the mid range tone in the shot. The card is held just in front of the subject (grey side towards the camera) to fill the field of view. It must be held absolutely upright so it does not receive more light on its surface than falls on the orchid flower. Ensure also that no shadows fall on its surface. Take your reading from the grey card and set your camera and lens accordingly. With automatic cameras, operate the exposure lock device to hold the settings, remove the card and expose the shot. With light coloured subjects you will have to give less exposure than the grey card reading indicates.

Whichever exposure method you choose, keep a record of your settings for particular lighting conditions and subjects and you will soon learn how to override your chosen system to get the best results.

Finally, take great care of your subject and other surrounding plants. Orchid seedlings may be growing around your chosen specimen, they can be tiny and thus easily crushed.

Camera equipment, including windbreak and reflectors positioned ready to photograph *Dactylorhiza fuchsii* subsp. *hebridensis* on Barra, 4 viii 1992.

FURTHER READING

Further Reading

The following list of references is by no means exhaustive. The literature on orchids is vast but some of it, like Darwin's classic book on pollination, van der Pijl & Dodson's work on the same subject and Dressler's *The Orchids* are all fascinating reading. A guide to other important works is *"A brief review of British and European Orchid Literature"* by J. J. Wood & P. J.Cribb, in *Natural History Book Reviews* 7: 17–21 (1984). The many Scottish local floras, both old and new, and *Watsonia* (the journal of the Botanical Society of the British Isles) are useful sources of information.

Angel, H. (1975). *Photographing Nature – Flowers.* Fountain Press.

Brooke, B. J. (1950). *The Wild Orchids of Britain.* Bodley Head.

Clapham, A. R., Tutin, T. G. & Moore, D. M. (1989). *Flora of the British Isles.* Cambridge University Press.

Darwin, Charles (1862). *Fertilisation of Orchids. The various contrivances by which orchids are fertilised by insects.* John Murray.

Davies, P., Davies, J. & Huxley, A. (1988). *Wild Orchids of Britain and Europe.* Chatto & Windus.

Downie, D. G. (1959). The mycorrhiza of *Orchis purpurella. Transactions of the Botanical Society of Edinburgh* 38: 16–29.

Dickson, J. H. (1991). *Wild plants of Glasgow.* Aberdeen University Press.

Dressler, R. L. (1981). *The Orchids. Natural History and Classification.* Harvard University Press.

Ettlinger, D. M. Turner. (1977). *British and Irish Orchids.* Macmillan.

Godfery, M. J. (1933). *Monograph and Iconograph of Native British Orchidaceae.* Cambridge University Press.

Hadley, G. (1970). Non-specificity of symbiotic infection in orchid mycorrhiza. *New Phytologist* 69: 1015–1023.

Harley, J. C. (1969). *The Biology of Mycorrhiza.* 2nd ed. Leonard Hills.

Hooker, W. J. (1821). *Flora Scotica, or a description of Scottish plants.* 2 volumes. London.

Kent, D. H. (1992). *List of Vascular Plants of the British Isles.* Botanical Society of the British Isles.

Lang, D. (1989). *Orchids of Britain.* Oxford University Press.

Lightfoot, J. (1777). *Flora Scotica.* 2 volumes. London.

Nelson, E. (1976). *Monographie und Ikonographie der Orchidaceen Gattung Dactylorhiza.* Speich.

Martin, W. Keble. (1982). *A New Concise British Flora.* Ebury Press.

Pijl, L. van der & Dodson, C. H. (1966). *Orchid Flowers: their pollination and evolution.* Coral Gables, University of Miami Press.

Stace, C. A. (ed.) (1975). *Hybridization and the Flora of the British Isles.* Academic Press.

Stace, C. A. (1991). *New Flora of the British Isles.* Cambridge University Press.

Summerhayes, V. S. (1968). *Wild Orchids of Britain.* 2nd edition. Collins.

Turrill, W. B. (1948–1971). *British Plant Life.* Collins.

Vermeulin, P. (1947). *Studies on Dactylorchids.* Schotanus & Jens.

Williams, J. G., Williams, A. E. & Arlott, N. A. (1978). *A Field Guide to the Orchids of Britain and Europe.* Collins.

GLOSSARY

Glossary

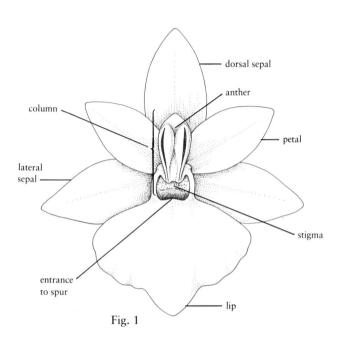

Fig. 1

Fig. 2

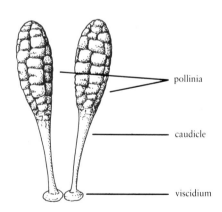

Fig. 3

acidic – water or soil with a pH value of less than 7.

alkaline (or *basic*) – water or soil with a pH value of more than 7.

anther – the male part of the flower which contains the pollen: in orchids this is at the apex of the *column* and is often referred to as the anther cap. (Figs 1, 4 & 11)

boss – a small but distinct bump in the middle of a flatter area. (Fig. 4)

bract – a small, leaf-like or membranous structure which subtends a flower; in the orchids this is usually linear or lanceolate. (Figs 4, 5 & 11)

bulbil – a small bulb (embryonic plant) produced on parent plant (on the leaf-tips of *Hammarbya*, the bog orchid).

bursicle – pouch-like area on the column between the stigma and the anther, covering the *viscidia*. (Fig. 10)

calcareous – rich in calcium carbonate.

capsule – a dry fruit which splits at maturity to release the seeds.

caudicle – the stalk attaching the *pollinium* to the *viscidium*. (Fig. 2)

column – the central structure of an orchid flower, formed by the fusion of the *style*, the *stigma* and the *stamens*. (Figs 1 & 10)

cotyledon – embryo seed-leaf contained within the seed and the first to emerge on germination.

elliptic – about twice as long as broad and tapering evenly to the rounded ends. (Fig. 3)

embryo – the undeveloped plant within the seed.

epichile – the front (apical) part of the complex lip of some orchids e.g. *Cephalanthera* and *Epipactis*. (Fig. 4)

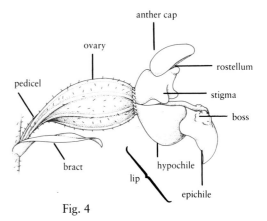

Fig. 4

epiphyte – a plant growing on another plant for support but not as a parasite.

glandular – bearing a usually spherical knob of secretory tissue.

hybrid – a plant derived as the result of the pollen of one plant fertilizing another plant of a different species (the result of the crossing of one species with a different species).

hyper-resupinate – term given to a flower which is twisted through 360°.

hypochile – the back (basal) part of the complex lip of some orchids, e.g. *Cephalanthera* and *Epipactis*. (Fig. 4)

inflorescence – the unit of flower or flowers borne on a common axis. (Fig. 5)

intergeneric – hybrid resulting from the crossing of two species of different genera.

interspecific – hybrid resulting from the crossing of two different species of the same genus.

lanceolate – three or four times as long as broad, tapering at each end but more gradually towards pointed tip. (Fig. 6)

linear – a uniformly narrow shape, more than four times as long as broad. (Fig. 7)

lip – the third petal of an orchid flower which is often larger and differently shaped from the other two and which, by a twisting of the flower stalk and ovary almost always takes up the lowermost position. (Figs 1, 4 & 11)

mycorrhiza – the symbiotic association between plant roots and fungi, vital for the growth of most orchid seedlings.

neutral – neither acidic nor alkaline, a pH of 7.

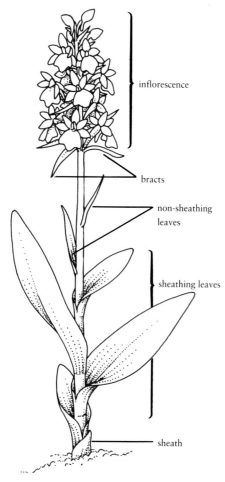

Fig. 5

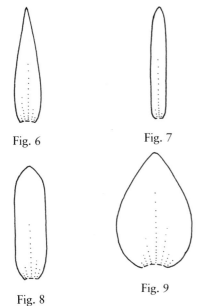

Fig. 6   Fig. 7

Fig. 8   Fig. 9

oblong – approximately two to five times longer than broad and with more or less parallel sides. (Fig. 8)

ovary – the part of the flower containing embryonic seeds (ovules) which develops into the fruit. In orchids it is always situated below the other flower parts i.e. inferior. (Figs 4 & 11)

ovate – egg-shaped in outline. (Fig. 9)

ovoid – an egg-shaped solid body.

pedicel – the stalk of a flower. (Fig. 4)

perennating – overwintering.

perianth – the sepals and petals.

petal – two of the three inner perianth parts, the third one being the lip. (Fig. 1)

pollinium (plural: *pollinia*) – more or less solid mass of pollen grains amalgamated together. (Figs 2 & 10)

protocorm – the minute tuber-like structure formed during the first stage of growth of an orchid seedling.

pseudobulb – a swollen, sometimes bulb-like stem.

reflexed – recurved, folded back.

rhizome – an underground, usually horizontal, stem which can give rise to a new stem at the tip and sometimes from the nodes.

rhomboid – diamond-shaped.

rostellum – the infertile third stigma of the orchid flower which lies between the functional stigmas and the pollinia. The *viscidia* are part of the rostellum. (Fig. 4)

saprophyte – a non-photosynthesising plant which obtains nutrition from soluble decayed plant or animal matter.

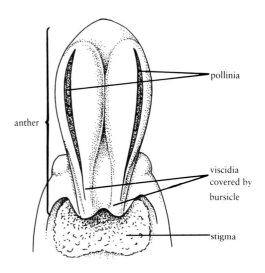

Fig. 10

sepal – one of the outer three perianth parts. (Fig. 1)

sheath – base of leaf which surrounds the stem, and can be either tubular or overlapping. The lower leaves usually consist entirely of sheath, with no leaf-blade developed. (Fig. 5)

spur – a hollow extension of the lip-base, often nectar-producing. (Fig. 11)

stamen – the male part of the flower.

stigma – in orchids the sticky, receptive area on the column which receives the pollinia during pollination. (Figs 4, 10 & 11)

style – the female part of the flower which connects the ovary and the stigma; in orchids forming part of the column.

tricornute – having three horn-like projections.

tuberoids – a tuber-like thickened root.

viscidium (plural *viscidia*) – sticky disc at base of pollinia which attaches them to a visiting insect. (Figs 2 & 10)

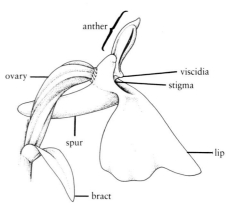

Fig. 11

INDEX

Index

Notes: plant names in italics refer to synonyms; page numbers in italics refer to illustrations